U.S. COUNTERINSURGENCY METHODS & THE GLOBAL WAR ON TERROR

An Assessment: Tactical Success and Strategic Blunder

U.S. COUNTERINSURGENCY METHODS ★★★★★ & ★★★★★ THE GLOBAL WAR ON TERROR

written by:

TERRY TUCKER

Tate Publishing & *Enterprises*

Published by Tate Publishing & Enterprises, LLC
127 E. Trade Center Terrace | Mustang, Oklahoma 73064 USA
1.888.361.9473 | www.tatepublishing.com

Tate Publishing is committed to excellence in the publishing industry. The company reflects the philosophy established by the founders, based on Psalm 68:11,
"The Lord gave the word and great was the company of those who published it."

Cover design by Elizabeth A. Mason
Interior design by Steven Jeffrey

Published in the United States of America

ISBN: 978-1-60462-256-0

1. History: Military: Strategy/United States

2. History: Military: Iraq Wars/Napoleonic Wars/Persian Gulf War/World War I/World War II

08.01.08

DEDICATION

I dedicate this book to my wife, Annette, whose support and encouragement has been unwavering. She has been my biggest fan, and without her encouragement and prodding this book would never have been possible. I dedicate this book to my two very young sons, Cody and Kyle; they have no idea what this is all about, but because of their mother's explanation are proud and supportive, I love you, boys! Lastly, I want to dedicate this book to my daughter, Jessica, my aunt and uncle, Aggie and John and my cousins Cindy and Jennifer and their husbands. Our geo-political conversations at our frequent family gatherings have been inspirational and enlightening, and provided the basis for some of my research.

ACKNOWLEDGEMENTS

While writing this book I have benefited greatly from the comments, advice and help of many people, some of whom I would like to acknowledge. I am indebted to Professor Charles E. White for providing a superior climate of learning and quality of education. Your questions, comments and encouragement were inspiring and invaluable. I also wish to acknowledge Dr. Geoffrey D.T. Shaw for his inspiring comments, critique, review and support. Thank you both for providing the most precious of commodities, your time, to helping me ensure that this project was more than just a success. I also wish to acknowledge a number of key people who are and were instrumental in my life, and either advertently or inadvertently acted as a source of inspiration for my research and writing. My wife, Annette, for your unwavering support and encouragement; my family; Major General Rick Lynch, US Army; Sergeant Major of the Army, Ret. Gene C. McKinney; Col Dave Collins US Army, Ret.; Col Matt Mattison, US Army, Ret.; LTC Chuck Garrison, US Army Ret; LTC Barney Morris, US Army, Ret.; Maj Dennis Bray, US Army, Ret.; Maj Alan G. Cody, US

Army, Ret.; and lastly, all of my friends and associates at the Kabul Military Training Center, Combined Security Transition Command, Training Assistance Group's IV, V, and VI—may God go with you all. Lastly and by no means the least, I wish to acknowledge Stacy Baker, Jesika Lay, Jennifer Bass, Lindsay B. Behrens and the entire Tate Publishing Company staff for their advice, support, encouragement and patience. Any errors in this work are all truly mine.

TABLE OF CONTENTS

FOREWORD

The United States and its coalition partners are now engaged in what is called the "New War," the war on terrorism. Since September 11, 2001, the US Armed Forces have been actively engaged in several simultaneous theaters of operation—Afghanistan, Indonesia and the war in Iraq.

Additionally, an important and critical distraction is the issue of North Korea and Iran as these countries continue to develop nuclear capability. Recent sparring incidents off the Korean peninsula include North Korean jets shadowing American spy planes in international waters, a US aircraft carrier off the coast of South Korea, and the deployment of strategic bombers to Guam. Other related incidents include the Israeli-Hezbollah conflict and the continuing fight to intercept insurgents breaching airline and airport security.

Clausewitz and Jomini might characterize these events as a "War of Opinion." Conflict based on ideological, religious and national passions. Certainly there is a growing Islamic faith and Islamic resistance to Western foreign policy and perceived hegemony. Some are referring to this as the inevitable clash of civilizations, a clash born of religious and ideological differences that have already been shaped and prepared in history.

There are references to a "Pax Americana," and 9/11 and the war in Iraq is the latest milestones in world history. Europe is dismayed and angered over what appears to be U.S. led aggression and delegates to the UN vow to veto resolution's calling for the use of force. Originally as of this writing, there were ninety-six countries world wide that had pledged military support and/or, have also provided support in the form of resources, diplomacy and economic assistance supporting the war on terror.

Since the events of September 11, a new doctrinal concept has appeared to have emerged that is used to describe this new war on terror. It's called *Asymmetrical Warfare*. However, this concept is not really new. In terms of recency, it appears that this concept first was mentioned in the 1997 National Security Strategy of which, it was also mentioned in the Quadrennial Review and QDR working group of the same year. Asymmetry was also used in the 1993 edition of FM 1, the Army, and in the 2001 version of FM 3–0, Operations.

There are a host of manuals that drive and guide the US military in its doctrinal concepts of waging war. For example, US Army Field Manual 3–0, Operations, is the keystone manual that proscribes war-fighting concepts. Additionally, there are a number of JP's or Joint Publications that also guide Coalition and Joint warfare.

Never the less, the doctrinal concept of Asymmetry *was not doctrinally* recognized until the publication of FM 3–0, Operations, and the latest edition of FM 1, The Army; although arguably, Sun Tzu and Liddell Hart also espoused the use of asymmetry in what they called the *indirect approach*. FM 3–0 and FM 1, give the concept of asymmetry short shrift. Perhaps at its worst it is lingoism that is an attempt to add a veil of mystery or an added level of complexity to modern war fighting and at its best, a new doctrinal concept. Never the less, there has been a lot of thought and energy devoted to this aspect of the art of war. Notably, the *McNair* papers provide a distinctly different definition of the concept when it is compared with the one in FM 3–0.

These two documents are both similar in how they both indicate that Armies can be compared in terms of symmetry and asymmetry; essentially, through the comparison of each force's current technology, organization, strategy and tactics. For the first time, U.S. doctrine, places a doctrinal emphasis and importance on the media and the intangibles of "psychological effect." We will follow this one step further and examine how *Disproportionate* effect, as one of the conditions of asymmetry is a doctrinal and strategic war fighting concept that has been around for a long time.

Modern examples include the German General Staff's strategic and doctrinal concept of the *Kesselschlacht,* which might be considered an attempt at asymmetry. Because technology, organization and arms were relatively speaking, equal, across most of the armies of Europe, the Germans relied on strategy and doctrinal components in order to shift the balance. Von Schliefen and Moltke expected the strategy to bring quick and decisive results. In fact by 1881, the plan envisioned defeating France within forty days and then immediately embarking troops for the Eastern Front.

Another example of strategic asymmetry can be taken from the so-called *Blitzkrieg*; a primary aim of the theory and practice of the *Blitzkrieg* was to turn a tactical advantage into a strategic one, albeit through the indirect approach using maneuver and firepower. In fact it is argued that maneuver is merely the platform that carried the firepower to the right place and time. There is no doubt that the German invasion of Poland and France using the concept of *Blitzkrieg* brought about a *Disproportionate Effect.* This approach might also be considered a form of attrition or even annihilation, but, essentially resulted in the destruction of weak units as the armored formations rampaged through rear area's causing immense physical and psychological disruption of units caught positionally and situationally unaware. This aspect of strategy turned doctrine was of a transient nature. Certainly mobility, speed and surprise could offset numerical advantage, but many other variables, soft variables, are also at work.

Understanding the need for an asymmetrical advantage is not hard.

Throughout history, the great captains have adapted strategy, tactics, techniques and technology in order to overcome their enemies. Frederick the Great was a "Maneuverist," Napoleon is considered the Father of the Corps organization and the "Maneouver sur les derriers;" the Germans would perfect the Kesselschlacht and Operational Art and the Americans and Russians would plan for the Deep Battle. Adding to the mix of elements and variables would be the partisan, guerrilla and terrorist.

Accordingly, does asymmetrical war change the doctrine of modern war theory? Do the current doctrinal principles of war change as well?

INTRODUCTION

I arrived in Kuwait in 2003, walked off the plane into the air conditioned terminal and was immediately struck by the heat; I shouldn't have been surprised, I have been to the Middle East before and was aware of how hot it could get. While I was attempting to adjust to jet lag I was given an introductory welcome briefing by CSM, retired, Steve Hicks, and the next day by my boss, Colonel, retired, John Shaw. My job would be operations planning and coordination. During my time in Kuwait I had many conversations with soldiers and officers that passed through our office. It didn't take long to realize that there was a disconnect between training and organization, yet, I couldn't quite put my finger on the issue. I could itemize a list of singular items but could not see through the proverbial fog or find the forest for all the trees. After this assignment I went to work in Saudi Arabia to write doctrine and advise the Saudi Arabian National Guard, and, then subsequently left for Afghanistan to write both doctrine and train the new Afghan National Army. It was during this period of doctrine writing and training in Saudi Arabia and Afghanistan that I began to realize that the challenges facing the US Army and its fight against the Global War on Terror were more than just singular items that transcended across all units, but rather, a larger dysfunction as a result of many components; that these macro components included Army doctrine, Army training and organization, and yet, it also included organizational culture, organizational psychology, institutional culture, and leadership. There was a strong realization that although the utility and use of force was a necessity in certain situations, there was also the paradox of a unit unable to shift its mind-set, decision-making and response's to a counterinsurgency environment. We continually applied decisive and overwhelming firepower, coupled with rapid mobility to put the right mix of firepower and resources at the right place to win the tactical fight, yet forgetting that in some case's it would have been better to do nothing at all. Increasingly, units are, and were forbidden to interact with the local population. Force protection measures took on a fortress like resemblance and the approach to counterinsurgency appeared to also be a 9 to 5 job. In specific instances units were advised by the unit intelligence officer (S-2), and instructed by the unit commander based on this advice that was coupled to a force protection risk assessment, that units would deliberately avoid patrolling specific area's because they were known to favor the insurgents and these areas were considered dangerous to US Forces because of that inclination. Units conducted operations separately from the coalition counterparts, did not coordinate with either coalition partners or even other US units and did not include the local police, security forces or army in its active patrolling. Increasingly it became clear that there was a huge disconnect between current doctrine, training, organization, methodologies and how to conduct a counterinsurgency and what the critical elements of success were. As I explored this issue further it was also evident that although some units and people knew what the doctrinal elements of a counterinsurgency were, the actual response was both contradictory and heavily kinetic. Essentially my very subjective findings were that some could talk the talk, but nobody could walk the walk. This was further reinforced as I had an opportunity to read the after action reviews from various units that wrote up there lessons learned for theater distribution. This was also important to doctrine writers so we could *evolve* and update the new techniques, tactics and procedures based on these new and evolving lessons learned. Lastly, as we wrote the updated doctrine and became engaged in teaching the Afghan Army it was increasingly apparent that the existing doctrine was convoluted and was spread out across several manuals that conflicted in definition and provided short shrift explanations for exceedingly complex theory, such as asymmetry.

In this book I am attempting to explain this huge disconnect between these various components that are both essential for how any army successfully fights, but also, how a successful counterinsurgency is premised on the tactical success of the small unit which in turn will deliver the larger strategic and operational success; yet this interrelated success of the tactical component with the strategic component is a product of tactical doctrine, tactical equipment, tactical organization, tactical training and formal institutional culture. It is clearly more than the promotion of learning and adapting to a bottom-up learning environment that inculcates and disseminates lessons learned. It is clearly more than fighting the inclination to wage conventional war against an insurgency. It is clearly more than decompressing or short cycling the learning curve to achieve rapid success. In fact, one might argue that without explicit descriptors and definitions, decompressing or short cycling, the learning curve might imply or infer short term ineffective work arounds with subsequent neglect for 2nd, 3rd, 4th and 5th order effects. Lastly, by direct implication it means that one requires more than direct knowledge of a myriad of non-military skills and that this direct knowledge of critical non-military skills is knowledge obtained from outside of the institution of military education.[1] Although I have accused the existing doctrine of using too many conventional examples to illustrate how asymmetry and counterinsurgency is explained and waged, one could easily accuse me of the same. Never the less, the examples I propose are also used in a manner to illustrate that the larger issue is one of doctrinal clarity and definition were either none exists or is completely lacking and misunderstood.

In order to begin our exploration, it will be necessary to begin with a few definitions. The following definitions are directly from current US Army Doctrine. *Doctrine is the concise expression of how Army forces contribute to unified action, campaigns, major operations, battles and engagements... Army doctrine also describes the Army's approach and contribution to full spectrum operations. As the Army's keystone manual FM 3–0 provides the principles for conducting operations. It describes the Army's operational-level role of linking tactical operations to strategic aims.*[2] *Doctrine represents a professional Army's collective thinking about how it intends to fight, train, equip and modernize*[3]

Accordingly, dogma as a synonym for doctrine is: assumption, concepts, credence, idea, postulate, principle, precept, tenet, theory, axiom, pronouncement, rule or universal law. It will be important to remember these definitions, specifically, in the sense of how the army conducts operations and linking tactical operations to strategic aims. With his in mind, it becomes readily apparent how complex the doctrinal process is and that doctrine cannot be developed in isolation.

At the risk of being accused of finding historical similarity where none may exist, it seems as though America and the US Army are in a similar situation as Prussia and the Prussian Army was from about 1866 to 1914. First, there was an interest in the theory and practice of war. The Franco-Prussian War of 1870 taught the major powers many lessons and they were anxious to solve the tactical conundrums. Similarly, we are also anxious to solve our own conundrum, albeit relearning the principles of counterinsurgency, but never the less, a renewed interest in the art and science of war. Second: Are we surrounded by enemies, albeit unseen? Third: does the military still prefer "doers," men of action, over the intellectual?

The current U.S. debate seems to promote a disbelief in theory. That we, as a profession, prefer a more quantifiable and analytical approach to war versus one that is infinitely more complex and appears to require more "soft" skills. I arrive at this conclusion less than scientifically through my own observations as an instructor, my interaction with Officers and NCO's in CFC-A, and, by surveying the professional reading lists of: The US Army War College, Chairman of the Joint Chief of Staff, the US Army Chief of Staff, the Command and General Staff College and the Combined Arms Research Library, which has also produced *Historical Bibliography Number 8*, Military Classics. Additionally, it must be noted that the Combined Arms Center Commander has a specific reading list dedicated to counterinsurgency.

Certainly there is a case that the importance of a diverse reading list is essential to a diverse and healthy grounding in the overall art and science of war, and admittedly, I have used and recommended these reading lists to many that have asked where to begin there journey in studying the art and science of war. Our development and understanding must begin somewhere. In essence, where and how does one begin to formulate one's ideas and theories about the conduct of war? How do future doctrine writers develop?

This link is tenuous. Determining and understanding the disconnect from tactical operations to the achievement of strategic aims takes someone from the "inside." An illustration of this might be found in the "Generals Revolt" and many that called for the resignation of Donald Rumsfeld. Only an insider that is familiar with the culture and organization understands the *soft* nuances that create a cognitive dissonance that resonates beyond passive-aggressive behavior. In many cases one might argue that the result of this revolt is more than just a clash of personalities. It is also argued and debated that the revolt is precisely over the spectrum of how to carefully shape, prepare and execute tactical operations to strategic aims and conversely, prioritizing the strategic aims in which one can begin to plan for the tactical operations that lead to achievement of these aims.

Despite the fact that the JMTC and the NTC are the vehicles used for training for war, the critical soft skills and local cultural nuances cannot be war-gamed, they must be learned and mentored. Units that go to these training centers are given a healthy dose of "generic" culture and scenarios that test their ability to work under pressure, plan missions in a stressful environment, and to use drills and SOP's as doctrinally written as a basis for evolving and adapting doctrine and mission tasks on the fly. In essence, the unit fights at the training center in accordance with the expected operational environment and in accordance with its prescribed mission. In turn the Mission Training Plan and Training and Evaluation Outline's for these units become the basis for pronouncing a unit combat ready or combat incapable.

To use a business phrase to clarify this counterinsurgency training, the Army is using the training center to train a unit how to *think globally and act locally.* But, thinking globally and acting locally is a useless proposition unless one has more than a generic understanding of the locations in which they will conduct operations. Fortune 50s call this process *due diligence;* Wal-Mart's pull out from Germany is a classic business case of not understanding the nuances of the local *Contemporary Operational Environment and performing due diligence.* Iraq is a classic military case of the same.

It is agreed that a unit can never be fully trained for the contemporary operational environment in which they may find themselves deployed, but is acceptance of doctrine as unwritten lawfulness and rigidity of action help one achieve the linking of tactical operations to strategic aims?[4] I am sure we know the answer to this question.

Does this imply or infer that the military has reached the point that the doctrine is false? Is it an indication that doctrinal myopia and single-mindedness have colored our ability to become adaptive? Is it cultural? Perhaps the answer to these questions is both a heavy dose of yes and no.

Let's explore this with an illustration. If we can make the assumption that current US Army doctrine is one that requires overwhelming and decisive operations, and that we seek to achieve some kind of disproportionate effect, then we must also assume that the premise of overwhelming and decisive operations is based on the principle of annihilation; annihilation that is either, or, both physically and/or psychologically overwhelming and decisive; decisive to the point of achieving strategic aims and decisive to the point that your adversary feels compelled, overwhelmingly, to abdicate. One might argue that a succession of operations of this type leads to the achievement of strategic aims. Conversely, how do military operations in counterinsurgency achieve decisive operations? The current US Military training paradox in counterinsurgency continues to measure success in terms of conventional historical examples by using illustrations such as: Jena-Auerstadt, the Prussian War of 1866, the Franco-Prussian War of 1870, the '67 War, and Gulf War of '91. Counterinsurgency operations seek disproportionate effect and decisiveness in a completely different manner; primarily through civil-social-government programs. This is not to say that military operations are not necessary, only that civil-social-government programs take the lead. Counterinsurgency is a constantly shifting balance of the right amount of security, application of force and social, diplomatic and economic programs.

If we examine the recent Israeli-Hezbollah conflict, the Israeli approach appeared to exactly mirror the US doctrinal approach of the Gulf War of '91. If we can assume that a tenet of decisive operations also includes the tenet of the battle of annihilation, that is, as somehow gaining space, while on the offensive, or gaining time while on the defensive, then as the attacker, I want to gain maximum space/territory with absolute minimum expenditure of time and resources. This is how I can defeat your decision cycle, by interfering with your time/space estimates, and disrupt your plan to achieve both physical and psychological annihilation (think of *Blitzkrieg* and disproportionate effect). As the defender, I want to trade the least amount of space for the maximum amount of time in an attempt to disrupt your attack planning. The defender disrupts the attackers decision cycle by holding on to territory longer than expected.

In the Israeli-Hezbollah war, the Hezbollah and its proxies appeared to have achieved both a decisive outcome psychologically and kept a reasonable amount of space for the maximum amount of time. But regardless of the actual outcome, it appears as though the battle of decisive operations has both a new timeline and requires new measures to link tactical operations with strategic aims!

This is the chimera that faces the US Army; an inability to adapt doctrinally, at least to the intangibles, in all aspects, is the crisis that we face for future asymmetrical warfare. One might possibly argue, based on overall performance during the last five years, that doctrine has become dogma. Another illustration that helps support this case is the US lead coalition in Afghanistan. American soldiers do not understand that their mere presence alone, the actual uniform itself, is a magnet for an array of perceptions both good and bad. Regardless of how well the US Army does, a school burning in the Eastern province or a *Night Letter* in Khost reaps exponentially adverse perceptions, your fault or not. US soldiers are a symbol, a representation of whatever the tribes, people or clans may attribute to it. Things that go well and all the good that is being done is an *Expectation because of your presence.* In other words the boundaries and/or demarcation lines between the US Army, the Coalition and NGO's is imperceptible to the local populace. In the minds of the *local*, either the presence or lack of social programs, medical care and security is all in the realm of the "green suiter." Either way, doctrinally, the US Army will need to become more associated and assist with NGO-type activities and/or, perform more actual NGO type work. How else to become seamless in doctrinal requirements for civil-military integration and what better way to learn the soft skills and attributes that are crucial to success? In other words our doctrine talks the talk, but does not know how to walk the walk.

In essence current U.S. doctrine touches on these combat multipliers and psychological aspects of war but does not encourage the training, practice or development of soft skills. It does not address "How To" be the *enabling agent.* Worse, is the fact that any experience gained outside of the "Army" as an institution is discounted and excluded as irrelevant and not appropriate. Additionally, the military passively-aggressively refuses to approach much of the required counterinsurgency training on any scale and prefers to rely on the random and haphazard affects of the ability of the individual soldier to be the "Ambassador" of both the United States and the U.S. Army. Arguably, many might say this is the realm of SOCOM and SOF and not the "Big Army"; but arguably, it is, especially if SASO and counterinsurgency operations continue to be subsets of Army Doctrine.

PROLOGUE TO CURRENT DOCTRINE

Political relations... had become so sensitive a nexus that no cannon could be fired in Europe without every Government feelings its interest affected...

Clausewitz, Book 8, page 590

Although current American doctrine is essentially Air-Land Battle based, military leaders conducting the counterinsurgency effort in Iraq and in Afghanistan recognize that some doctrinal shifts need to be made. For instance; The Center for Army Lessons Learned regularly publishes CALL notes and bulletins that utilize experience gained from the major training centers and from operations with in the active Theaters. Examples of this include updating techniques, tactics and procedures for Cordon and Search, Sensitive Site Exploitation, and Clearing a Cave Complex. [5]

In each of these cases, the updates focus on disparity and/or discrepancies in the existing drill manual that prescribe precise actions for Team to Battalion-size units conducting these types of operations, or develop completely new drills where none previously existed.[6] The intended purpose of which is to fill gaps and holes and to counter enemy measures that have adapted counter measures to US Army methods. In essence, the current *evolving doctrine* for Battalion and below, and, Company and below is updating its ability to adapt to tactics, techniques, and procedures by updating tactical doctrine drills. The positive aspect of this is that tactical lessons learned improve survivability of both the individuals and also the small unit. However, how are joint and coalition lessons learned at the Unit of Action (Brigade Level) being updated and disseminated? Furthermore, how is overall Capstone Doctrine evolving and updated?[7] For all the academic debate within the NDU and Strategic Studies Institute, the only real doctrinal change as of this writing seems to have been the task organization and designation of Units of Action and Units of Employment.[8] A seminar sponsored by the U.S. Army Combat Arms Institute appears to reinforce the notion that despite the idea that transformation is a key subject, and counterinsurgency a key element in that transformation, that they still had one eye focused on conventional operations.

Although FM 1, *The Army*, addresses the fact that fundamental paradigm shifts are being made throughout the military from organization to doctrine, it will take an immense amount of effort to insure success. Is this really transformation or is this just modernization? We will not understand the impact, nor, will we know the answer to this question for at least one full generational cycle as the military attempts to change culture, mindsets, institutional training and organizations in order to achieve the *operational flexibility and adaptability*... and make the military... *culturally aware... sensitive to differences and the implications those differences have on the operational environment.*[9]

PRELUDE AND A SHORT DOCTRINAL SURVEY

The history of US Army doctrinal development can be shortened to a couple of key eras; the development of airmobility operations during Vietnam, the post-Vietnam development of Air-Land Battle doctrine and the post 9/11 era of the Global War on Terror.

The US Army defines the gestation of airmobility as the mid 1950s. A tactical doctrine manual, FM 57–35, "Army Transport Aviation-Combat Operations" was written, and a provisional *sky cav* platoon was formed, which essentially, through extensive experimentation, eventually became the nucleus of the 792d Aerial Combat Reconnaissance Company (Provisional).[10] One of the first official steps to transformation occurred on 15 January 1960 with the formation of the Rodgers Board. The Rodgers Board made several recommendations regarding helicopter type, design, funding and policy. One of its most important recommendations was the recommendation to "prepare an in depth study to determine *whether the concept of air fighting units was practical* and if an experimental unit should be activated to test its feasibility."[11] Although the scope of review of the Rodgers Board was limited, it provided the beginning's of essential guidance for development and procurement and was indicative of the vision of transformation that was in its embryonic stage. [12] As airmobility experimentation proceeded, the Howze Board was officially appointed on 25 April 1962.[13] The board had an extremely demanding schedule, and was required to submit its final report by 24 August 1962. General Howze was given wide latitude in which to convene the board to include dealing directly with DoD, the Department of the Army, other military services, government agencies, civilian industry and to convene the board at other installations as he saw fit.[14]

The Howze Board final report was submitted on 20 August 1962, four days ahead of schedule, and the formation of the "air assault division was the principal tactical innovation."[15] Although many recognized that change was essential, the recommendations from the Rodgers report, the interim field testing and subsequent Howze Board Report were that the Army would enhance combat effectiveness in both conventional and counter-guerrilla actions *and could accomplish other tasks with smaller forces in shorter campaigns.*[16] The new tactical innovations—by inference and implication—were supposed to support operational and strategic objectives, and were focused on combining all the elements of combat power, maneuver forces, reconnaissance, communications and service support. The formation of the air assault division and the air cavalry combat brigade was to combine the classical functions of cavalry operations with the air assault division's role of closing with and destroying the enemy on the ground. In essence the new innovations provided the mobility to move maneuver forces quickly, and provide organic and immediate fire power through aerial weapons platforms combined with fixed wing assets to provide and perform the traditional indirect fire role that had previously been dominated by the Artillery and Air Forces close air support roles. Additionally, the aerial weapons platforms provided enhanced standoff, and support by fire positions from the realm of the second dimension—the air. In hindsight, although these tactical innovations provided an enormous amount of mobility and inherent firepower, they also contributed to the idea that tasks and campaigns could be effectively shorter and or shortened, and again, by implication, the speed of which maneuver delivers superior firepower would be the capstone of doctrine.[17] Although the innovation of airmobility was to enhance the counter-guerrilla operations in Vietnam, the development of doctrine proceeded along conventional thinking, and the doctrine of airmobility received its baptism by fire in the Ia Drang Valley at LZ X-RAY when the 1st Bn 7th Cav, 1st CD conducted air assault operations into the area with the specific mission of Search and Destroy Operations.[18] LTC Moore would change his tactics, techniques and procedures slightly based on the intelligence estimate he received, and directed that all units use the same landing zone instead of separate landing zones for each company. Further, the timing and synchronization of the air assault with the artillery was timed to within H -1 minute, not much room for error.

The United States Army would continue to emphasize that airmobility doctrine was nothing short of a doctrine that was a subset of conventional warfighting, albeit adapted to the Vietnam War. Additionally, US Army *Capstone Doctrine* essentially remained unchanged from the 1941 version of FM 100–5; and airmobility doctrine helped to further reinforce the notions of short sharp campaigns of short duration that were characterized by overwhelming firepower and an increase in mobility. It is extremely hard not to draw the correlation or similarity that American airmobility doctrine was looking at ways to emulate and exceed the "gold standard" of maneuver warfare in the integration of firepower and mobility such as the German experience of WWII, or more specifically, the German campaigns in France and the Low Countries in 1940. Nevertheless, US Army doctrine would experience its next major change ten years later in 1976.

General William E. DePuy assumed command of the newly formed Training and Doctrine Command in 1973. He would be the impetus for changing US Army doctrine and essentially established the foundation for the current Air land Battle doctrine. TRADOC worked hard for the next three years developing the new doctrine. The doctrine reflected the times and conditions the US Army was facing at the time. The US was slowly beginning its withdrawal from Vietnam, major US policy shifts had occurred, the defense budget was on the decline and the Cold War seemed to be escalating; all these elements worked to influence the authors of the new manual. In 1976 the revamped FM 100–5, Operations, was released and it immediately caused a controversy that led to the manuals displacement—an event that was never anticipated nor intended. Despite the fact that General DePuy worked hard to define the complexities of doctrinal change and establish the importance of doctrine, General DePuy was unable to overcome institutional resistance and was unable to instill confidence in the new doctrine.[19]

In 1973 US policy makers began to consider there options and came to the conclusion that American ability to contain, deter and provide conflict control was limited and the administration would have to establish priorities in accordance with Strategic interests. As a result, American priorities shifted to Europe/NATO with a nervous eye on the Middle East; In terms of conflict control, assistance and other help, Africa, Asia and everyone else would have to fend for themselves now. This new strategy was called "The Strategy of Realistic Deterrence."[20]

As planners re-assessed the situation it became apparent that while the US was embroiled in Vietnam, the Russians were busy upgrading their posture in Europe. Since 1965 the Russians had added five new tank divisions, deployed a number of Warsaw Pact units closer to the border, upgraded T-54s, T-55s and other equipment, and had fielded the T-72. To US planners all this was patently obvious; the Soviets not only had an increased offensive capability, they were planning on a non-nuclear preemptive strategy. Additionally, Soviet willingness to intervene in the 1973 Arab-Israeli War also demonstrated a disconcerting level of assertiveness. Considering the extremely sorry state of physical and moral readiness of the US Army in 1973, it is easy to see how hypersensitive US policy makers would have been to all these elements.[21]

FM 100–5 was a single doctrinal articulation of how the US Army would conduct offensive and defensive operations. Various editions of FM 100–5 released during the period 1944 to 1961 presented few and only subtle changes. For example, the US Army Pentomic Organizational structure change represented a huge doctrinal change and change in organization; however, it was never incorporated into the revised editions of FM 100–5.

The 1976 edition of FM 100–5 stood the Army on its head. This was the first manual that would be a "capstone" manual for an entire family of manuals and would result in the wholesale replacement of the US Army tactical doctrine. The new manual represented a new and overarching concept in which other proponent branches and supporting arms would have to follow. *This was the first manual that made an attempt to not just change the organization of the US Army, but attempted to change the way the entire US Army would think.* This was the first manual to describe "how to fight" and was a complete break from the past. It emphasized the fact that the Army must fight outnumbered and win, and it must win the first battle of the next war, which was, and is not, a US Army historical tradition. The new manual would emphasize that the tank was the decisive ground weapon but it could not survive unless it was part of a combined arms operation that included tactical air. The new FM was both singular and visionary in its approach. Despite the fact that it was Europe/NATO focused, it articulated the concept of the forward active defense, advocated no reserve, and first coined the phrase "Air-Land Battle," in which the US Army recognized the interdependence of US Army and US Air Force assets, and was a first attempt at unifying the battle space and battle command. Although the resultant doctrine was not entirely new and was an amalgamation of Clausewitz, German lessons from WWII and the '73 war, it was prescient in its prediction of a new lethality, improved technology and fluid, more mobile operations.

In some respects the doctrine postulated by General DePuy was not really that new. The concept of the active forward defense was a popular German doctrine of WW II. It had been perfected on both the Eastern and Western fronts. The doctrine postulated that a commander should place all his units forward, trade as much space for time as possible, and to laterally move units from the left and or right to combat the main assault, meanwhile accepting some level of risk elsewhere. Explicit and implied in this concept was the idea that a commander had to perform a very healthy intelligence preparation of the battlefield, and these subsequent positions would have to be pre-identified and pre-prepared. This meant that many primary, alternate and subsequent positions would have to be identified in each battalion, brigade or division sector. It also meant identifying specific units for assault and pursuit if the opportunity to seize the initiative occurred. Additionally, German concepts of suppression and support by fire placed a heavy emphasis on direct fire weapons; those fires were more accurate and responsive because they were under the command of the local commander. Creative commanders could add, delete or adhoc any number of weapons systems together in order to form a support by fire element—based on the immediate situation, or developing situation—in which to either defend or attack. Crucial to the offensive nature of this new doctrine, was this concept of "overwatch," in which tanks and infantry formed a team and "overwatched" the area in which the assault would occur. In this way, the overwatch element could provide immediate and accurate direct suppressive fire on any position while the assaulting team could maneuver to the flank or rear of its immediate objective area. Another critical element in this new doctrine was the specified and implied task of effectively using every fold in the terrain and the effective use of camouflage.

The seeds of the 1976 version of FM 100–5 were sown in the lessons of WWII, merged with the technology and lessons of the fluid and furious pace of the Yom Kippur War, and tempered by the fiscal and policy constraints in place in 1973. Not only would the Army have to think differently, it was also going to have to be trained differently—from the ground up. Success in the embryonic Air-Land Battle doctrine would rely on a number of key principles and tenets such as speed, mobility, agility, surprise and concentration at the right time and place coupled with a systems approach to weapons capabilities and weapon systems integration. It also placed a very heavy emphasis on quality over quantity.

In the introduction to *The Roots of Strategy, Book 4*, David Jablonsky references the concept of thinking in "Time Streams" as "The core attribute for such thinking is to imagine the future as it may be when it becomes the past."[22] Thinking in this fashion should provide a basis for thinking in terms of action—reaction—counter action; simulation exercises in thinking that should help determine the strengths and/or weaknesses in our asymmetrical planning, and more importantly, the basis for challenging assumptions as we write evolving doctrine.

FM 1, *The Army,* states that doctrine is an authoritative statement about how military forces conduct operations and provides for a common lexicon with which to describe these operations, and that doctrine is the basis for army curricula. The manual further states that, above all, the army will continue to fight and win its nation's wars. This is further reinforced in FM 3–0, *Operations* that:

As the Army's keystone operations manual, FM 3–0 provides the principles for conducting operations. It describes the Army's operational-level role of linking tactical operations to strategic aims and how Army forces conduct operations in unified action. FM 3–0 bridges Army and Joint operations doctrine. It also links Army operations doctrine with Army tactical doctrine.[23]

Additionally, FM 1, *The Army*, also describes doctrine as "the concise expression of how Army forces contribute to campaigns, major operations, battles, and engagements. It is a guide to action, not hard and fast rules. Doctrine provides a common frame of reference across the Army."[24] Although these two keystone manuals taken together as key doctrinal statements may provide a definition of what doctrine is, and is intended to serve, paradoxically current doctrine and training is still heavy on conventional methods and pays lip service to the "Soft Side" (less coercive methods) of Security and Stability Operations/COIN (SASO/Counterinsurgency) doctrine. My direct observations and involvement within CSTC-A, TAG IV (Combined Security and Transition Command-Afghanistan, Training Assistance Group IV) is that doctrine, strategy, organization and training are still too heavily tied to conventional methods. At the various meetings and training sessions it is agreed that we are fighting a counterinsurgency, but the supporting conceptual umbrella of overlapping non-military tasks of Security and Stability Operations (SASO) are not considered crucial tasks to successful war fighting. Even as we update current Tactics, Techniques, and Procedures (TTPs) at the tactical level, there is clearly no method to incorporate these lessons into operational and strategic doctrine and there is less willingness to minimize the "conventionalness" of operations or to recognize that the soft side of war is crucial and integral to success.[25] The army still seems reluctant to accept its role in SASO/ Counterinsurgency, even after five years of war, and seems to regard the idea that this kind of job belongs to someone else. The implication is that this someone else is Special Operations Command (SOCOM) and SOF Teams.

Implicit and inferred in the Army's definition of doctrine and operations is that training, task organization and success at winning wars is a holistic concept with each interdependent on each other for overall success. Implicitly, if one element is missing or flawed then the presumption is that failure can follow.

If we can assume that FM 3–0, *Operations* is the capstone for execution, and military force is the decisive component of that execution, then the coupling of strategy from the National Security Strategy through the National Military Strategy down to the actions of the squad and team are crucial, linked and interconnected.[26] The paradox (or schizophrenia) of American doctrine can be seen in the timeline and release of selected capstone manuals. For instance, FM 100–20, *Military Operations in Low Intensity Conflict* was released in December 1990. It wasn't updated again until October 2003 and replaced with FM 3–07.22, Counterinsurgency, an interim manual that expires October 2006; and FM 3–24, *Counterinsurgency* was officially released December 15 amid much whispering about the content and continues to draw public attention[27]; the differences in all three of these manuals is striking.

For instance, FM 3–07.22, in the introduction of the manual sets the foundation for confusion by indicating, "the army has conducted stability operations which also include counter-insurgency operations."[28] In comparison with FM 3–07, *Stability Operations and Support Operations*; a manual considered *principle doctrine*, and although it directly speaks to counterinsurgency, albeit sparingly and in a brief section of page 5–6 and Appendix D, counterinsurgency is considered a major subset *of* the overall principle doctrine of Stability *Operations*. On the other hand, FM 3–24, counterinsurgency indicates that stability operations are a sub-set of its principles and tenets. It is clear that stability operations are important, but to what degree are they, or do they become the primary doctrinal principles? In other words; what, or is the degree of separation between counterinsurgency and stability operations? At what point does the National Command Authority decide that stability operations are no longer applicable and that it has evolved to counterinsurgency? When does a counterinsurgency devolve into SASO? What is the key trigger points that signal the end of one and the beginning of the other? By simply naming the operation so, or when other interests dictate?

Paragraph 1–47, FM 3–07, *Stability Operations and Support Operations,* almost sums up the focus of the entire manual when it devotes one short, sharp paragraph to *decisive operations.*[29] This is a crucial point because current military culture emphasizes that military operations in all aspects must be *decisively won and decisively executed.*[30] How do you execute a decisive SASO operation when it is inferred that SASO is more relationship and cooperation based than conflict and coercive based? How does one achieve disproportionate effect as one of the conditions of asymmetry?

Other major faults include its continuous reference to FM 3–0, *Operations*, which rehash's the military decision making process, tenets of army operations and the principles of war. All in all not a bad thing, except that the current Military Decision Making Process (MDMP) is conventional in its orientation, and along with its key supporting actor, Intelligence Preparation of the Battlefield (IPB), needs to be drastically tailored to suit counterinsurgency doctrine and operations. Lastly, FM 3–07.22, *Counterinsurgency*, by implication, appears to have superseded the overall conventional focus contained with the two principle doctrinal manuals of FM 3–0, *Operations*, and FM 3–07, *Stability and Support Oper-*

ations; this sets the stage for the existing doctrinal conundrum. The current revolution in military affairs and subsequent transformation efforts have been less than transparent other than the implementation of a new modular design task force (the new old Battle Group?); and no new capstone doctrine has been developed either. Once again, despite the tactical lessons learned and the subsequent updating of tactical doctrine at the platoon and company level, how will these lessons be incorporated into army wide capstone doctrine? Are the current definitions of the principles of war and the development and execution of strategy at the operational level still valid? Why do we cover SASO and counterinsurgency in seven different manuals that are not consistent in the how they convey the basic doctrinal message?

Adding to the doctrinal chimera is the fact that the US military has recently rediscovered counterinsurgency warfare, and not coincidently, there are diverging ideas on why we may or may not have neglected the doctrine of the small war.[31] Nevertheless, the problem persists, doctrinally, how to adequately define and set parameters for this kind of war and from the historians view, attempt to seek continuity in the past (as prologue).[32] A perusal of any of the military manuals referenced so far, and one will find historical examples throughout the materials to support and defend current thoughts on conventional doctrinal precepts.

Seeking these past continuities, ideas and challenges has seemed to have added to the doctrinal debate in the US Army's attempt to find historical connections where none may exist. If we can assume that theory grounded in historical example and analyzed through academic debate leads to the precepts of doctrinal development; then even if the theory and doctrine are flawed, it has played a key role in shaping the experience and the continuing debate over how that experience should be interpreted. Regardless of how we arrive at our interpretations, the issue now becomes one of confirming legitimacy on a framework when that framework becomes written as doctrine. Does this now become a self-fulfilling prophecy? If there are innovations in doctrinal thinking that are contrary to current institutional thought, how does one overcome the institutional and cultural pressures?

In essence, the military has done a fine job of analyzing the literature and writing tactical level doctrine that is commensurate with its definition of what doctrine is and means to the US Army.[33] All of the manuals address how to conduct and/or defeat an insurgency albeit through Operations Other than War, Security and Stability Operations, Low Intensity Conflict and Counterinsurgency, but[34] it has not addressed what it intends to do to achieve the cultural, maturational and experiential requirements to meet the political, economic, diplomatic and social demands to achieve the necessary synchronization of skill sets, training and institutional culture/training to match success with doctrine. My personal conversations with an SOF Field Grade officer simply confirmed this when he indicated, " I know what I want and need in a guy when I see it."[35]

Critically, the issues cited above are directly related to the US Army's ability to adapt to the current operating environment; regarding this, two major essays circulating within the US Army academic and professional arena both explicitly and implicitly support these notions. In one instance the issue relates to an ability to adapt strategically and the other in its ability to adapt operationally and tactically.[36] Explicit and implicit in the US Army's ability to adapt is the fact that doctrine, counterinsurgency and strategy are interrelated and interdependent and if there is a disconnect at any level it will most assuredly effect the other elements. With this in mind, when is it appropriate to discuss the implications of tactical failure and how that failure leads to operational and strategic failure?[37]

Parallel to this line of thought is the US Army's approach to doctrine and operations in Afghanistan. Currently, the US Army is using Embedded Training Teams (ETTs); teams that are from three to sixteen men and consist of a combination of active, guard and coalition units that are assigned to Afghanistan National Army (ANA) units at various levels from Corp to Battalion. The mission of these ETTs is to train current US doctrine and to advise the ANA during combat operations (the inference and implication is that the ETT are *doctrinal experts*). The success of the ANA is predicated on the same principles that the US Army is predicated on—a holistic interdependent concept of doctrine, training, operations and organization. In essence, we are betting the success of the ANA on both the knowledge gained within formal institution training such as Basic Training or the Staff Operations Course, and the knowledge and experience that these ETTs impart coupled with performing actual missions. Regrettably the range of abilities of, and within the ETTs is markedly inconsistent. They are fundamentally weak in many of the area's that they are supposed to be training the ANA on.[38] As of the time of this writing, General Durbin, CSTC-A Commander was meeting with key leaders to discuss the relevance of adapting doctrine for the ANA. Unofficial comments indicate that General Durbin would like to see two or three page doctrinal TTP to short cycle the doctrine staffing process. This short cycle thought process potentially makes several assumptions:

1. That doctrine is established, read, understood and practiced

2. Institutional and unit training is based on a current established doctrine and assumes a learning/change cycle that is approved by the chain of command.

3. There is a continual unit training and improvement cycle

that not only reinforces existing doctrine but also implements and integrates the evolving doctrine.

It is my observation of units from three ANA Corps that not all of these assumptions are valid and doctrine is neither understood nor practiced by the ANA. Additionally, not all of the ETTs are up to the current task of training and mentoring doctrine for the ANA, and MPRI, Defense Contractor has the explicit mission to train the ANA *and the implied task to mentor the ETT's.*[39]

If we can assume that doctrine is a reflection of the times and tempered by the contemporary fiscal and policy constraints of the day; it is also a reflection of the experiences and lessons learned. General DePuy's strategic assessment of the '73 war, coupled with American interpretation of lessons learned from WW II seemed to re-inforce his notions and concepts regarding tactics and training. Key elements that insure the success of doctrine, this relationship of doctrine, implied procurement strategy tied to doctrine, training, organization and supporting tactics, techniques and procedures is all highly complex and dynamic. If technology changes too rapidly to what extent must doctrine change to adapt to the technological change? Conversely, if the research, development and acquisition process is a long and arduous process, then to what extent is the procurement process affected and doctrinal change commensurately stalled due to inappropriate and or outdated weapons systems? Some modern theorist(s) might argue that this is a modern basis for asymmetrical warfare.

During the period of US Army doctrinal development during, and post-Vietnam, as well as today, doctrine developers collected lessons learned, wrote academic literature and fielded tactics, techniques and procedures which did not carry the official weight of doctrine, but did provide the baseline for comments from the field in which doctrine could be based and revised. Assembling and producing this literature in the form of TTPs and call notes directly and indirectly influences training, procurement and the development of future doctrine. In the case of General DePuy, he charged his staff with rewriting *both* the capstone manual and all the proponent manuals in eighteen months. Conversely, the US Army has been at the GWOT for five years now and has failed to produce a single revised capstone document[40]. It has produced a wealth of tactical lessons learned and translated these lessons to TTPs, handouts, circulars and other tactical literature, but has failed to translate the evolving tactical success to strategic doctrine. A partial reason for this seems to be the fact that there is a plethora of information and literature on insurgency and counterinsurgency and the Army's inability to consider the context of the social-cultural context of the Afghan environment.

Some of the best-written literature on counterinsurgency seems to have originated in the 1960s, specifically, David Galula's, *Theory and Practice of Counterinsurgency* and *Pacification in Algeria 1956–1958*. Recent literature from the Combat Studies Institute, The War College and the Center for Army Lessons Learned have promoted and released a host of essays on the same subject, yet consistently refer back to David Galula and T.E. Lawrence, or modern theorists such as Ian Beckett, Bruce Hoffmann, Colin S. Gray and Bard O'Neill. There does not seem to be shortage of available literature and lessons learned, so why has the Army failed to adapt tactical lessons into an updated doctrine and training program that ties the importance of non-kinetic tactical success with long term strategic success?

Based on my observations at both the field and institution level there seems to be an inconsistency in knowledge about counterinsurgency. Additionally, there is less agreement or consensus on how to actually conduct operations and even less agreement on how to formulate doctrine in support of that performance especially where it concerns the application of soft, non-kinetic and SASO techniques.[41]

OF ORGANIZATION AND DOCTRINE

A short survey of American Divisional and Brigade task organization reveals that today's basic organization has remained essentially the same since 1941.[42] Of course there have been some minor adjustments with some additions and deletions, but the basic Brigade level organization has mostly consisted of three tank battalions, three infantry battalions or a mix of infantry and armor; and it has consisted of maintenance and service support elements, some C2 and reconnaissance elements. Engineer, Artillery and other service and support elements have traditionally been assigned to division. [43] In the 1980s, with the inclusion of joint operations, battalion and brigade elements received assets from division, traded companies with other battalions and generally developed a "habitual" relationship based on mission assignments. Attached and detached units normally, with some very few exceptions, went and worked with the same brigade or battalion that they had always been cross attached to. Increasingly, Brigade combat teams that trained at the NTC or JRTC went with "their" slice of divisional assets so they could be a combined arms team for the purpose of mission, training and deployment. With the army's configuration of the SBCT and the new modular design, a lot of the traditional assets that were assigned to the Divisional level have been reduced in size and assigned to the new UA's. For instance an Infantry division may have had a combat engineer battalion; the new UA now has an engineer company and heavier engineer assets are assigned to the UA "Y"or Corps level.[44] In many respects the new modular design is representative of past task organization for training and deployment would, or might have normally looked like. It appears as though the traditional divisional staff and headquarters configuration has been eliminated in favor of a Division "enhanced". But the reduction or elimination of the Division Headquarters is superficial as the command and control remains in place.

Accordingly, the new organization is supposed to represent an increase in deployability and flexibility to commanders SASO Mission; however, a look at the new organization actually reveals nothing new, and still places a heavy emphasis on conventional operations. The new Brigade Combat Teams (BCTs) vary in size from three thousand to four thousand soldiers, and are supposed to deploy anywhere in the world in ninety-six hours. Accordingly, if the new Stryker Brigade Combat Team/modular organization is a reflection of both evolving and proposed future doctrine then the emphasis has shifted in favor of deployment with one weary eye on conventional operations. The deadly chimera in this self-fulfilling prophecy of organization change to match doctrine is that the new modular design still infers and implicitly places a heavy emphasis on current US versions of the principles of war and war fighting. A specific example is this one of several statements that the Army used to rationalize its force realignment initiative [to] Divest itself of Cold War structure to better fight the war on terrorism. The Army intends to capitalize on joint force air superiority and the demonstrated capability of joint precision munitions by decreasing the number of field artillery, armor, and air defense units. [45]

But is this new structure really better able to fight the war on terror? This statement seems to support the tenet of "offensive" and appears to substitute the principle of "mass" for the principle of "precision." Additionally this new organization appears to imply radical organizational shape-shifting in order to react to the *conventional* needs of SASO and counterinsurgency. But what of the mental and cultural shape-shifting that also needs to take place? What does one do about the counter-intuitiveness of the conventional orientation to SASO/counterinsurgency in current doctrine? What about the penchant for a doctrinal one size fits all response and the realities of SASO and counterinsurgency? Although the US Army denies the fact that it fights wars of attrition, what do we call the patient use of precision munitions to take out High Value Targets? Regrettably, what *appears* implicit in the new organization is that the conventional doctrine and organization are still immutable and SASO and the GWOT mission belong to someone else other than the Army.

Although the new organization may relive some stress in terms of units with a high demand for deployment; without understanding what the support UA's for the new organization look like we will not know if the UA or UE "E" or "Y" will have the assets required to perform the SASO operations such as Psyops or Civil Affairs units. Even the new UA's are configured for what current doctrine calls decisive operations, pays lip service to SASO/Counterinsurgency and barely tolerates the need for Special Operations Forces.[46]

Relying heavily on the work of Colin S. Gray, war and warfare is dominated by context. Despite the fact that we are addressing doctrine, the revolution in military change or transformation is an important subset of doctrine. As the Army addresses change in the midst of conflict, the inference and implication is that RMA, doctrine, counterinsurgency and its host of related peripheral elements should be regarded in a holistic manner, but regrettably, the internal social-cultural environment of the US Army precludes, inhibits and denies the integration of this process. Although some senior officers within the military recognize the importance of changing the culture, many from the Pentagon down to the platoon leaders level are resistant and don't see a need to change.[47] Additionally, some are arguing that the US is unsuited for a *long war*. That this unsuitability is more than just mismatched doctrine and execution, but is also a mindset and other attributes that are considered serious deficiencies in approaching the long war in a successful and holistic manner.[48]

In essence, our mandate of finding and implementing new and innovative operational concepts and adaptive evolving doctrine means that changes in organization, technology, force employment and doctrine that are not considered to have a decisive enough effect will not be given much—if any—consideration at all.

If we can assume that counterinsurgency is a skill set that needs its own level of core competency, then the debate and improvements in technology and organization within the US Army over the last five years has been fundamentally centered on improving existing old conventional core competencies. It has relied extensively on a "frame of reference" rooted in the victory of 1991. Despite the network-centric capabilities, and advanced technology and implementation of the modular design, the required change in mindset that is required for success is currently counterintuitive within the current US Army leadership. As Colin S. Gray has so aptly expressed "... there can be more than one contemporary military enlightenment." [49]

The conventional military of today shows a remarkable awareness, yet is blatantly dismissive of anything that challenges current mindset and doctrine. The militaries dismissiveness is a product born of institutional bias, culture, victory, technology, arrogance and zero tolerance. More than the idea that arrogance and conviction have been born from victory in the Gulf War and the conventional phase of the Iraq war, how can these past victories now suggest that we need to radically shift thought and doctrine 180 degrees to a new direction? Additionally, we have a deeply-rooted problem with our doctrine, one that is driven by bias, ethos, values, culture, victory and conviction; all the components that make up the psycho-social values and mores of a large institution. U S Military doctrine is predicated on decisive victory with overwhelming force. Even the current manuals on SASO and counterinsurgency place an emphasis on decisive victory. Decisiveness stated and implied in current doctrine is derived from concentrating firepower, speed, agility and maneuver, which in turn is predicated on reliable technology, and effective lethality is considered the gold standard for how one achieves decisive victory. Decisive victory is the standard in which all operations are analyzed for and against, which is overwhelming force rooted in precision and copious amounts of firepower. Will that be enough to influence the perception, hearts and minds of an asymmetrical enemy?

APPLICABLE LESSONS FROM CONVENTIONAL WAR

In the official histories written after World War II, heavy criticism was aimed at the American strategy for the war in Europe.[50] The official histories speak volumes to the coalition planning process and the Army seems to be especially anxious to vindicate itself. [51] Although it is important to discuss and learn the value of strategic, operational and tactical lessons, it is equally important to understand and discuss the political and attitudinal barriers that caused military planners and strategists to execute less than ideal plans. Especially when, in some cases, the military planner seems to have a greater appreciation of the impact of potential political consequences and a greater astuteness of the impact of global conflict than those charged with the political leadership of the country.[52]

It has been said that the allied strategy of coalition warfare in WWII was the search for common denominators among the three primary allied players: Britain, America and Russia.[53] But was it really the search and compromise of these denominators or was it perhaps, more appropriately, the question of how to implement the necessary military means within the boundaries of stricter political and attitudinal forces? How can military planners effectively suggest and implement a strategy when the popular constituency does not necessarily agree with the body politic? Today, as in our past, the same philosophical arguments cause much hand wringing and finger pointing. It is widely known that Roosevelt had tempered his road to war decisions against the expectations of the American constituency. More importantly, American military planners were cognizant of a host of military weaknesses as a result of many years of neglect born of political and popular support that was in opposition to strategic necessity. As a result, regardless that military planners and staff had identified a host of possible options beginning back in 1934, the reality of the present state of military readiness and popular opinion were the primary drivers of what that strategy would actually be.[54] Those are the barriers to planning and implementation that military leaders and planners still face today.

It was generally accepted that the British had the most experience in war, diplomacy and coalitions, that the Americans were naïve, and the Russians inept and incapable. It is also generally agreed that the British approach to war was one of the "indirect approach";[55] a nibbling at the edges, using all the elements of national power to balance any perceived or actual increase in threat from Europe, but this had always been British policy. Additionally, British officers were not only used to working and planning in conjunction with their political leadership; British officers were also experienced in "governing" their provinces and (with some alacrity or sluggishness) would administer decrees within their authority. British military governors had wide decision-making latitude. It is not within the scope of this paper to discuss success or failure as a result of those decisions; it is important to note that British officers had a one up on their contemporaries in the realm of political-military interaction and planning.

Similarly, the American view was one of support by proxy, but that view was born from a policy of Isolationism. Like the British indirect approach, America was willing to provide material, economic and diplomatic support in order to get as least involved as possible.[56] Both Britain and America realized the necessity of this indirect approach; however they each came to different conclusions on why it was necessary. Additionally, for different reasons, British and American officers applied planning and strategy techniques in order to achieve its ends, but from completely differing political and attitudinal differences. For Britain it was the logical extension to its political philosophy and its means to managing the empire. On the other hand, America arrived at the indirect approach conclusion primarily for reasons that appeased national attitude. It was necessary because of fifty years of prevailing American popular and political attitude that had only been reinforced by our subsequent interpretation of the events of WWI and the interwar period. It was necessary because the military recognized that it did not have the ability or means to execute any other strategy, at least for the time being. For Britain it was "The Primary" strategy. Both Britain and America were attempting to achieve attrition, albeit slightly differently, but primarily for different past, present and long-term reasons.[57] In that regard, *ante bellum*, America could have cared less who the power broker in Europe was so long as it did not affect them. On the other hand, Britain wanted to maintain the *status quo ante bellum*. Although both America and Britain were fully aware of the carnage that WWI had inflicted; America had escaped that war relatively unscathed in comparison with the major powers. As a consequence, the indirect approach to the Allied strategy of WWII had completely differing and definitional differences between them. The consequent and sometimes acrimony between British and American military staffs was difficult to comprehend by the Americans because of its lack of experience in global politics. [58]

In many respects, the British and American aims were similar; give the necessary material, economic and diplomatic support to further their aims by proxy. In America's case it was a result of a myriad of past decisions, and past and prevailing attitudes that created the conditions resulting in the one and only feasible and immediate conclusion that both military and civilian planners could agree on. On the other hand, for Britain it was a premeditated extension of its long term and historic policy of war, diplomacy and empire management. A recent example of such support by proxy would include American support of the insurgents during the Russian invasion of Afghanistan

America's idea of security is that it is inherited; that our oceans would provide the foundation of future policy and security decisions, and for a very long time this attitude was not only sufficient it was also evident in the decisions that the government made in terms of size and composition of the force. The Army and Navy planners had to compensate for the past neglect and impact in the series of "Rainbow" plans that the US Army developed prior to WWII. Despite the fact that the initial defense line for the Western hemisphere changed several times, the plans do not seriously address anything other than a forward Naval presence that would bear the brunt of the defense, and a few Army divisions for a potential invasion of the West Coast that none thought would happen anyway. America's biggest concern was with a large population of Japanese on the West Coast whose patriotism and reliability was seriously questioned. In many respects the Joint Army-Navy war planners, although politically astute, were forced to recommend less than ideal strategies because of political decisions made between 1898 and just prior to the war. Staff planners recognized the economic and military importance of the Philippines and all the Islands. They made recommendations in 1903 that would have placed us in a more favorable position. Those recommendations were supplanted by political and popular attitudes that disagreed with those charged with maintaining our security. As a consequence, the pre-war predictions of the early loss of the Philippines and all the major islands were realized and initial and subsequent strategy would have to trade space and time for human blood. War Planners knew all this in 1937 but were powerless to improve the situation.[59] Strategy and planning had to compensate for the realities of the present. It does not seem hard to imagine how frustrated the planners would have felt knowing that a stopgap strategy was the best they could hope for. Not only in terms of a "Europe First" strategy, but also in terms of an inadequate defense posture that would require tremendous effort on the part of those doing the defending. The sad thing is, planners knew that all the garrisons were going to be written off as early as 1936. On the other hand it is not hard to imagine how frustrated the British were with American politicians and war planners' short sightedness in terms of the after effect of military action and its impact on politics and diplomacy.

MITIGATING THE LESSONS

The military and its institutions are influenced, fashioned and created by two things. First; by the values and ideologies of our democratic institutions; and secondly by those elements the military perceives to be threats to the State and its interests. The perpetual conflict between military strategy and politics is the result of, on one hand, minimizing present and future threats to the State and its interests; and on the other hand, an attempt to balance its social requirements and obligations. In other words, the maximum amount of security at the least cost and sacrifice of its values and to its democratic and social institutions.

In all cases the military has pretty much ran its wars the way the politicians have wanted them run. On the other hand, the American public has not always agreed with its governing body, the military or both. And therein lays a paradox between policy and strategy. It is the nature of the beast that the military must continually look forward to address all the potential threats to the existence of the state. It is also a requirement to make strategy recommendations based on history, trends, and potential present and future capabilities of all potential opponents. The essential job of the military and its strategy, simplistically, is to anticipate wars and then win those wars. The other merging variable is the fact that internal security was not really assumed to be a subset of strategy and policy; although the Guard and Reserve might be required for occasional disaster assistance and/or riot control, internal security belonged to the myriad police and law enforcement agencies. This all changed as of 9/11. The basic root of the problem is why do we chastise the military for planning and implementing strategy under strategic conditions that it was not allowed to help shape?[60] In most cases, political and economic decisions that created those strategic conditions over a period of thirty years completely disregarded military recommendation that would have enhanced both the military and political/diplomatic posture.[61] Certainly the military predicted certain general events, but the overall orientation, structure and composition of strategy is a result of what the politicians and America wanted it's military to do. In other words, prior to 9/11 would the US Government and the general population have condoned anything remotely resembling the Patriot Act? Additionally, what about a military "pre-emptive" or "preventative war" strategy like the one implemented in Iraq?

By implication, military strategy implies turning over the whole war making mess to those that are the professional warrior's. It means by extension, inference and overt acceptance that the military has the backing of the government and more importantly, the people. But the illusion of strategy is that it is influenced by many variables, the most important of which are the political, social and societal norms and expectations that fit within an acceptable framework of what that strategy should not only look like, but how it will be resourced and executed; and on the other hand by a core of military and civilian leadership with is own diverging and yet vaguely similar institutional and social norms and expectations. Not the least of which is the belief that molds almost all strategy, that the strategy and war should or will be a conventional one with a heavy over reliance on national and historical tradition which was molded on past history, values and formal and informal institutional norms.

DEFINING ASYMMETRY

Arguably, a historical review of America at war has led many to the conclusion that there is an American way of war, but some argue that we do not necessarily have a way of war, but a way of fighting battles.[62] History seems to support the notions that we have a distinct way of fighting—strategies and fighting that seem to emphasize technology and attrition. But attrition, annihilation and exploitation now seem to have to be supplanted by pacification, or at the very least unarmed neutrality; in essence a larger and greater subordination of military action to political action. But Clausewitz has always told us that this is the case: by extension war and various degrees of conflict intensity are really an *extension and proxy of politics and just as importantly, guerrilla tactics (read that as asymmetrical war)* is an adjunct to war.[63]

Accordingly, Clausewitz tells us that "War is an act of force to compel our enemy to do our will," but equally important is the fact that "Attached to force are certain self imposed, imperceptible limitations... and that the maximum use of force is in no way compatible with the simultaneous use of the intellect. . . . . passion of hatred cannot be conceived as existing without hostile intent. . . . .the slightest nuances often decide between the different use of force. . . . we can now see that in war many roads lead to success, and that they do not all involve the opponents outright defeat. . . . . the choice depends on circumstances. . . . one could call them arguments ad hominem." [64] So what are the principles of asymmetrical warfare? Are the principles of asymmetrical warfare different from conventional war or limited war?

Despite the fact that Clausewitz is better known for his contributions to the philosophy of war in a conventional sense, he was also acutely aware of the variables that we would consider asymmetrical. If we assume that Clausewitz is correct and that the moral aspects of war are as important as the other principles of war, then it may be easy to see how Americans will draw the wrong kinds of conclusions regarding success or failure for the current war on terrorism.[65] McNair Paper 62 addresses in detail a high-level view of the origins of asymmetrical warfare and also proposes a definition of its own after establishing what some of the prevailing definitions of asymmetrical warfare are. [66] The author proposes, the existing definitions, while narrowly accurate, seem insufficient in explaining asymmetry. It will be argued . . . that a better definition of asymmetric warfare would be: *Leveraging inferior tactical or operational strength against American vulnerabilities to achieve disproportionate effect with the aim of undermining American will in order to achieve the asymmetric actor's strategic objectives.* The key differences in this proposed definition are the element of *disproportionate effect*—achieving strategic objectives through application of modest resources—and the explicit recognition of the importance of the *psychological* component. These elements are essential to considering how an asymmetric actor can achieve strategic objectives through an operation—even a failed operation—

that, from the perspective of the larger power, is only a tactical attack."

The distinction is important, especially for a war that cannot show immediate and visible success, but is entirely contingent on a series of multi and simultaneously successive combination of elements, principles and variables that may not show results for years to come, and the results it does produce may not be readily recognizable or worse, produce other unintended consequences. Equally challenging is the realization that Americans have a collectively short attention span, and will not tolerate any notion of an inability to quantitatively compare the fruits of success with the demise of an opponent. Americans, like the Austrians of the Napoleonic era will demand compensation and satisfaction of some kind. [67] With the above in mind, how did we manage to wage conflict successfully in the past? What was the benchmark that validated that success? Assuming that it is either the complete destruction of the enemy's force or subduing the enemy by other means, what element or elements contributed or achieved that *disproportionate effect*? In other words, considering past historical examples of any kind from any era, is it not safe to assume that a commander has always attempted to take advantage of any weakness and achieve disproportionate effect from any element at any level?

CONTEMPORARY WAR

The evolution of US Army war-fighting doctrine can trace its roots to 1939. However, the 1976 version of Field Manual 100–5 set a doctrinal precedent and articulated a totally new doctrine for defense, the active defense. The concepts and principles articulated in the 1976 version evolved into what became known as Air-Land Battle, and became formal doctrine in the August 1982 version of FM 100–5. The doctrine was revised several times over the years and was considered validated in the Gulf War of 1990–1991. Fm 100–5 was replaced by FM 3–0 in 2001, and this new FM is considered a reflection of current Air-Land Battle, joint, and operational doctrine that is the continuing progression and transition of its 1939 roots.[68]

If the military is at fault for executing a flawed strategy, then that flawed execution must begin with the lack of intentional planning. Some have argued that the nature and parameters that form the basis for strategy, parameters and bounds that are stated by political leadership, remains frustratingly unclear and subsequently, by implication, military planners are unable to formulate the military means to achieve political ends.[69] In essence, how does a commander know which doctrinal elements to use and/or adapt if he is unsure of what the ultimate aims are? Does he default to *decisive operations; or* does he attempt to combine a set of ROE that may inhibit how those doctrinal principles are to be applied?

The products of modern technology have not changed the way strategy is viewed. Although joint operations are emphasized, the emphasis is primarily to maximize the advantage and depth of weapon systems along multiple planes on multiple dimensions of the battle space. In other words, how does one combine land, air and naval systems to maximum decisive effect in the battle space from near to deep and from land to space? On one hand, it appears that current strategy is multi-level and viewed dimensionally; planning is still linear and ignores the more important effects of human and intuitive skills. Thus, strategists still think in terms of trade-off between alternative options of military force in an either/or analysis of the use of force. But that is the modern paradox of military strategic planning; it is not just necessarily the distinction between the different types of light or heavy units of infantry employment or types of aircraft, but rather who is best equipped with the skill sets to find, fix and counter social asymmetric differences. This distinction becomes clearer when we understand that current US military thought on strategy planning focuses on identifying centers of gravity.[70] In essence, current doctrine implies that defeating operational and tactical forces comes when one presumably follows the principles of war and defeats one's center of gravity,[71] and that the delivery of one very swift and decisive blow will result in immediate victory. As an example, the US Army attack on Baghdad in the Iraq war. With this in mind, is current doctrinal thought and practice tied too closely to the tactical and operational elements of war? How does one plan for an asymmetric war? Or do contemporary illustrations of centers of gravity rely too heavily on the wrong definitions of how to define a center of gravity?[72] I would suggest that we have applied the wrong definitions for an asymmetrical war and that future planning requires a major shift in the notion that we might be better served if we emphasize economic, ideological and diplomatic effects-based planning as opposed to capabilities based planning, and that planning considers the potential for multiple centers of gravity.[73]

It is as much about competence in warfighting and unequivocally knowing weapons platforms capabilities and weaknesses as it is about understanding human sensibilities and translating that to operational/tactical action with long-term positive effect. How can speed of information, loads of integrated data, remote sensors, digitized units, enhanced lethality and smaller operating units make for better strategic plans? By themselves, they cannot. The chimera of modern planning and technology is that technology has increased our level of information, but we have not increased our awareness or understanding of what that information means and we have traded decentralized execution for centralized execution at the highest levels. More importantly, technology has finally given us the tools in which to execute the spirit and intent of von Moltke's *Auftragstactic,* yet increasingly we appear to neglect the necessary *bildung,* and appear to have minimized decision making and autonomy at the lowest levels; levels that the original intent of the reformation in technology was intended to benefit.[74] In essence, the evolution of US Army doctrine has been this emphasis of tying *auftragstactic* to operational centers of gravity and expecting immediate strategic decisive results from the tactical and operational elements. Once again, witness the conduct of operations as the US Army conducts it last phased attacks in and around Baghdad.

Some have argued that the framework of a strategy consists of five elements: context, objective, capabilities, assumptions and costs. Others might say that it consists of six questions without answers.[75]

1. What is [this war] all about? What specific national interests and policy objectives are to be served by the proposed military action? How great is the value attached to those interests and objectives and what is there fair price?

2. Is the national military strategy tailored to meet the national political objectives? In other words are the means and ends matched to meet those objectives?

3. What are the limits of military power? In counterinsurgency more power can translate to less success.

4. What are the alternatives?

5. How strong is the home front? Does public opinion support the war and the military strategy used to fight it? How much stress can civilian society endure under the pressures of war? Is the war morally acceptable? Can it be justified as a "just war"?

6. Does today's strategy overlook points of difference and exaggerate points of likeness between past and present? Has concern over past success and failures developed into a neurotic fixation that blinds the strategist to changed circumstances requiring new and different responses?[76]

In any case, modern Army manuals such as FM 5–0, *Army Planning and Orders Production,* also provide a methodology for analysis, but also emphasizes the physical, the scientific and the quantifiable. Paradoxically, one method of strategy—Boyd's—focuses on the art and intellectual side of strategy, but is not widely accepted within the military.[77] Boyd does not provide a neat linear checklist, but rather advises gaining an appreciation for such things as: self-interest, critical differences of opinion, internal contradictions, frictions, and obsessions.[78] Using Boyd's advice as a guide, if we want at least one clear example of how a failed thought process contributed to a failed strategy and in turn a failed execution, then one only needs to look at the pattern of events in the Middle East. Additionally, the Israeli forces do not appear to be either containing or deterring the number of both incidents and attacks on its civilians. [79] In all cases the analysis and subsequent planning phases that led to the formulation of strategy neglected the analysis and impact of how to come to terms with defeat, presence, ideological change and the inevitable forced and imported cultural imprinting that comes with military occupation and reconstruction.

There is a triad of manuals that essentially comprise the corner stone of Army Doctrine; yet the key manual on Operations, FM 3–0, sets the foundations for schizophrenic behavior within the first eight paragraphs.[80] While espousing the nature of *full-spectrum operations* and acknowledging the changing face of battle, it emphasizes OOTW and liberally applies a number of vignettes to support this while attempting to show the interdependence on the strategic, operational and tactical levels. Through out the entire manual, one is confused about its intended specific audience and orientation. If this manual's intent is to provide the doctrine of military operations for the Army, its fault is that it does not know if on one hand if it is aiming to teach those at the Battalion Level or Divisional and Corp Level. With one eye on OOTW, its underlying focus is on conventional operations. Its discussion of the framework of operations indicates that versatility and agility are maximized in order to develop multiple capabilities within a unit, yet the clear indication is that this does not include units below Brigade level. This is ironic when one ponders the reasons for our technological, doctrinal and organizational advancement over the last thirty years as we have advanced the original Air-Land Battle doctrine. Once again, the original intent was to put the information and decision-making tools at the tank turret level. Additionally, all previous thought and training was (and is) on how to utilize auftragstactic to defeat operational and tactical centers of gravity that would in turn lead to strategic success with a strong inference on achieving overwhelming disproportionate effect with the success.

It would appear that FM-3 and current lessons learned in the GWOT would continue to emphasize a methodical battle strategy that fights battles not wars; one in which we are still struggling with how to keep from losing the element of tactical and operational surprise through a conventional shaping of the battlefield with smart and conventional weapons. Although the intent of the smart technology was to reduce collateral damage and improve surgical targeting, we still face the same dilemma of the effects on strategy that leaders did in WW II. It is common knowledge that MacArthur had the strictest bombing policy anywhere. His message to all his commanders emphasized that Filipinos "will not understand liberation if it is accompanied by indiscriminate destruction of their homes, their possessions, their civilization and there lives." MacArthur approved the target list. [81] He knew the implications to the overall strategy at all levels despite best American efforts to minimize unintended damage and consequences. Currently in the war in Iraq, American commanders seem to have miscalculated both the media and Iraqi response to alleged American missteps in precision targeting. As a result, many still question the viability and success of the strategic, operational and tactical plans.

The basis for understanding and translating strategy into military options and operations must first be understood in terms of the relationship between political and military action. The center of gravity for translating the political-military aim is no longer at any level above Brigade. It now lies between every size unit from the individual soldier, squad and battalion. The Brigade and higher echelon units are now just the element, which procures, obtains, and allocates the necessary resources for these small units.

This is the paradox that must be resolved within US Army Doctrine. How to reconcile the concepts of auftragstactic and centers of gravity and decide how to integrate the strategic with the operational and tactical elements of war and planning, realizing that the units at and below Brigade or Unit of Action level now require a level of skill and knowledge in domains normally outside of the traditional military knowledge base, such as politics and business.

Politicians provide the highest place in the hierarchy by providing the political aims, aims that change and compete with other changing aims as the general population and/or the media demand. Understanding and prioritizing these demands falls within the realm of the political-military leadership that is charged with formulating strategy, however, the real understanding and execution must occur at the small unit level. Long-term success is not just determined in the Pentagon or the White House, but also by the radio operator and rifleman on the streets of Baghdad. Technology is such that time and space have been compressed to the point that senior leaders have no time to digest, think about and react to change. The intuitive and human skills and means employed or used by small units to achieve political/military success will ultimately decide how successful American foreign policy really is.

The entire decision-making cycle is supremely condensed. If we want a higher probability of success, then the decision process will have to be taught and understood at the squad level, or the entire organization needs to be re-thought in terms of level of experience and skill within the Unit of Action and the Unit of Deployment.

In that case should majors be company commanders? Should captains be platoon leaders, similar to the British model?

Let's use the premise of current doctrine to explore this idea briefly. If we accept that current doctrine is Air-Land Battle based, and that the original basis for our doctrinal foundation is based on the German concepts of *Schwerpunkt*, or critical point, and, *auftragstactic*, or mission type orders, then we must also accept the fact that this doctrine is based on a level of trust, competence, skill and ability at the lowest possible level. Remembering that the original German concept of maneuver warfare, from which our doctrine is derived, is based on the idea that tactics and operational doctrine was to combine techniques, trust and a certain level of education at the tank commander level in order to achieve the decisive result. Examples to support this illustration include the German *Stosstruppen* tactics of WW I, German armored units in WW II, and Israeli tank and company commanders in 1973.

Is this premise of a higher level of skill and maturity at lower organizational levels highly unrealistic? In any event, the point has been driven and it needs serious consideration. The question remains: In order for the strategy to be successful, what level of experience and skill do we require at the small unit level and should doctrine be rewritten to reflect this? If special operations and joint force are the tools for the model for a successful strategy, how much of the conventional forces that assisted these operations either contributed to, or hindered, success of the strategy? The reports from Operation Anaconda in March 2002 in Afghanistan would lead one to believe that despite the use of elite regular Army units used in conjunction with Special Operations operatives, those regular Army units may have contributed to the failure of that mission.[82] What skill sets and experience do we need at the small unit level to achieve a higher level of competence in the conventional force when conventional forces are supposed to supplement special and joint operations achievement of the political, military, pacification, and hearts and minds campaign?

How will political means be achieved when it appears that we continue to say that strategy is no longer linear, yet we continue down linear planning paths? Distressing is the fact that media and technology access has appeared to politicize the officer corps to the point that it's completely acceptable for senior officers or their staff to leak information,[83] especially when there is a divergence (bordering on acrimony and enmity) between civilian and military leaders on not only the strategy, but the nuances of the strategy that normally fall strictly within the realm of military planning. Additionally, both the media and planners must also accept some blame for allowing pictures and Op-eds to take the place of facts. The media portrays its interpretation, foments and incites riots within the debating circles with its carefully crafted opinion, and uses pictures to project perception to further its agenda and political inclination. In turn, planners and leaders react to the perception of public disagreement or agreement.

In the past, strategy raced with time in order to overcome the effects of disease, desertion and climatic effect. In many ways, technology has almost eliminated these as primary variables in strategic planning. The primary variable now is to compete with the changing attitudes of coalitions, politicians, and the general population. The center of gravity, of course, is no longer a capital or an economic center that will guarantee collapse. In many respects the real center of gravity has been, and always will be, the opinion of the popular vote. Nevertheless, the culminating point in battle is no longer a specific battalion in action or battle; increasingly, the culminating point that will win or defeat a strategic plan is the result of many simultaneous culminating points in terms of ideology, the symbolic centers of that ideology, multi levels of economic support, and many transient people that supplant the "capital". One only needs to look at the three primary factions within Iraq—Sunni, Shiite and Kurd, coupled with the myriad factions within these that give one an idea of the dimensions of complexity. This does not even address the tribal and clan loyalties within the same multiple levels of factions. Has our over reliance on technology and speed supplanted the human and intuitive skills that are also the basis for a successful strategy? Or perhaps more importantly, do our neglected human and intuitive skills affect our ability to adapt to rapidly changing events on many planes and dimensions that impact our strategy? Is the implementation of replicable process templates our new strategy? Our new evolving doctrine?

A US Army online publication, "On Point", appears to validate that the strategy for Iraq and subsequent War on Terror was a conventional strategy that relied on the last ten years of improving joint, network-centric, technology-oriented conventional war.

Although the Army has generally counted on the number of available combat brigades in its planning scenarios for the last twenty years, it was doctrinally understood that the Division was the primary C4ISR element that processed, directed and controlled activities and resources. Lessons learned in the last three years seem to indicate that this may have been reduced to the brigade level and that SOF is the glue that made the joint ops successful.[84]

The current strategy for Iraq seems to have focused exclusively on military action, and does not give much thought to the short and long-term political consequences of the action. All the lessons learned to date lean heavily on conventional strategy. Although this is a departure from Cold War and previous strategy, it is realignment towards concepts that rely heavily on speed and transmission of information. In essence, a shortening of the cycle of target identification to target engagement, not because the methods used in the last fifty years were necessarily all wrong, but that technology has made the old processes and organizations obsolete. Changes to warfare have continued to condense and compress process and organization, which in turn have increasingly made Clausewitz philosophy on war more appropriate for modern multi-level dimensioning and complexity. In other words, what is the new cycle time for learning how to fail fast? Because of the immediate news effect how does one recover? It would appear that technology has exponentially increased the amount of time in a recover period after a blunder. On the other hand, one might hope that either complacency or media desensitization would result in outright apathy.

If we can assume that major warfare has seen less real change than we now suppose, is it safe to assume that the future holds far more doctrinal continuity than we now expect?[85] If the modern system of warfare is based on balance estimates and that current and future international policy rests on the assumptions of those estimates, then the threat of war and the pursuit of victory will in all probability continue to be capabilities based. Additionally, should we presuppose that the modern system of war has not seen a significant change in doctrine in approximately the last two hundred years, and that success is predicated on lethality, should we also presuppose that increased lethality over other key elements is the key to future success? Current doctrine, which is Air-Land Battle based implies that current doctrinal methods of maneuver warfare are actually a synonym for actual movement and physical placement, which in turn is based on capabilities-based assumptions. If asymmetrical warfare is the antithesis of current doctrine then how do we learn to adapt to the seemingly increased chaos and friction of war planning that seems to defy capabilities based planning and assumptions?

The lessons learned and the implemented changes seem to validate that contemporary American strategy is battle oriented, and has a heavy reliance on technology-driven, multi-dimensional attrition to the exclusion of human elements of strategy planning.

As history has shown, technology developments have greatly condensed the many variables of war. The current war in Iraq seems to demonstrate that units below brigade are now the key element in strategy planning. Throughout history, the size of organizations for strategic planning, counting the numbers available for a planned military action, has gone from the Army, to the Corps, to the Division. In WW II, the leadership gambled on a ninety-division limit. Today we count the number of Brigades or Units of Action.[86]

Again, with what appears to be an emphasis on winning each individual battle within the area of operations, the failure seems to have been how to subsequently and seamlessly link both the military response with, and simultaneous to, the political, societal, economic, psyop and reconstruction response. Not only does it appear that we failed to adequately plan a holistic strategy and response, more importantly we have failed to communicate not just the strategy, but the individual and collective successes to date that would add substance to the perceived ambiguity of the present strategy on the GWOT. If there is a new trump card in strategy, the war of ideas, and the manipulation of ideology and mass perception through the media is that wild card. In the past, the defensive pause gave one the opportunity to rebuild and reorganize. In the future there can be no defensive pause. Even if the pause seems to be a military necessity, the pause must be masked with deception, and the media tool is how both the initiative and offensive is maintained. Future strategy will increasingly be decided by the perceptions of general populations and international organizations, perceptions as understood and believed from what has been portrayed by the media. Maneuver warfare, positional advantage and freedom of maneuver, are, and will be, dictated by real time media reports. As a result, all actions and counter actions need to be viewed and thought of in advance in terms of what the world response will be and demand. If there is a new element for strategic analysis in planning, it is the elements of regional and local mass psychology, not necessarily to effect a massive and immediate ideological change in behavior, but rather the subtle and slow manipulation of ideological change through doubt and what outwardly will appear to be logical and plausible credibility.

Asymmetrical war brings unprecedented challenges and questions to the military planner, questions such as:

- Where and how does asymmetrical war now provide the power of decision?
- Is a decisive result achievable, and what is the focus required to achieve that result?
- Is tactical decisiveness the principle instrument of contemporary (read asymmetrical) battle?
- How and does asymmetry change the quantum relationship of force multipliers?
- How does one effectively utilize the new Battlefield Operating Systems (BOS) of media, coalitions, consensus, culture, economy and diplomacy?
- How does the planner achieve complementary and reinforcing effects in civil affairs and economic planning
- How does one quantify the soft power and cost-benefit ratio of military intervention?
- Why does planning have to focus on the fewest possible sources? Wasn't the Air-Land Battle concept designed to focus on multiple simultaneous sources?

As planners attempt to transcend conventional planning modes, serious conventional planning questions will demand unconventional and coordinated responses from untraditional staff areas and resources.

As we continue to analyze modern war in terms of asymmetry, it is important to note that current definitions of asymmetry address the idea that strategy, tactics and the very understanding of western paradigms in prosecuting war has changed, but the nature of war itself has not changed.[87] Because warfare, all warfare, is based on social, economic, technological and political elements, the new mitigating factor that changes the essence of the management of violence in a so-called asymmetrical war is contained within the media and in soft diplomatic and interpersonal skills. In other words, we must tangentially shift to results-based planning that also uses the term "maneuver warfare" to include planning outside of traditional military elements and contexts.

Real time media reporting and availability of information has greatly compressed time, space and decision making. Furthermore, the potential impact that decisions formerly had on operations at the strategic level were always analyzed in terms of impact on the strategic and political aspects of war beginning with a Corps-size element and higher, whereas it is now registered and felt at the individual and squad level. In other words, the skill sets, decision making and leadership has essentially escalated down. Recent examples of the impact of this ripple effect of individual actions on the overall perception of America's ability to prosecute the war include the Abu Ghraib prison scandal, the Jessica Lynch story, and the recent revelations of the 82nd Airborne of more prisoner abuse.

The additional complications of business and globalization complicate the planning sessions because American plans are not intentional in terms of long-range impact, generally fluctuate every four years with a new administration and show absolutely no level of patience or tolerance for anything that resembles compromise and certainly nothing less than the prospect of total victory. To illustrate this example, the North Vietnamese were willing to use all manners of any means to achieve its ends, including the fact that they might have to regress a step or two; that success is incremental and victory may be more than twenty years in the future.[88] Assuming that Americans now acknowledge that post-modern warfare (warfare post 9/11) includes a larger social-political element of war and that this element is largely managed (possibly manipulated) by the media, to what end can we reduce the paradox of impatience that Westerners have increasingly incorporated into the culture?

We are frozen in the past in terms of how to plan for future success. Our own inability to change manage expectations of others and what those benchmarks of success might look like is a significant factor in a perpetuating bogus circular logic that effects our decisions around our foreign policy, economic policy, and military planning. To illustrate this, look at the Gulf War in 1990–91 and the Somalia debacle of 1993. It is these types of incongruence in the political-military dynamic, global and/or regional apathy and a host of other factors that contribute to a less than effective military operation that create asymmetrical opportunities. The accumulating revelations of prisoner abuse and a lack of training of CSS Units (such as with the Jessica Lynch story) reported globally in real time also shake the very foundations of both the military and political strategy, and have far-reaching, long-term effects on foreign policy that are not easily repaired

Although the perception is that the American way of war is "more a way of battle than an actual way of war",[89] the credit for this error does not rest entirely with the military, but with a succession of administrations that were driven by popular consensus, and (right, wrong, or indifferent) other political considerations. Once again look at Gulf War 1990–91, Somalia, Kosovo and Bosnia. Equally important is the idea that our technology and network-centric operations give us tons of information, but are we transforming that information to useful knowledge and intentional decision making for an effective long-term hearts and minds campaign? The fabric of military decision making consists of gathering effective and timely information, but network technology and systems has effectively gathered and stored more information than can possibly be used, and the errors in judgment is now the inability to decide what is the correct information to use, and result in analysis paralysis, haste in decision making, and how to use that information effectively with a human/political touch. Increasingly, technology has given us the tools, yet we have exponentially decreased our human analytical and interaction skills in order to rely exclusively on technology applications.

Although some might suggest that military planners develop solutions to problems that they prefer to solve, this is not entirely true.[90] For instance, it is now widely known that General Schwartzkopf reinitiated new planning guidance for a potential war with Iraq several months before the Iraqi invasion of Kuwait in 1990, contrary to Pentagon plans or the administration's guidance. In essence, it was a commander's initiative looking past contemporary narrowness. The critical success factors that define military success are not necessarily defined by the military, but also by political and economic criteria as well. Success in the first Gulf War was judged by the amount of destruction rendered on the Iraqi Army. If other factors would have been taken into consideration then the invasion of Iraq in 2003 might not have been required; this lends credence to the idea that shortsighted analysis is short-term prevention.

Continuing political, economic, diplomatic and international legal interpretations will continue to add incongruence to political-military tenets. This incongruence is the weak link that asymmetrical fighters will exploit, and this continuing fundamental pattern of incongruence will disrupt both military plans anc ability to effectively use force. The easily identifiable transitions o political-military dynamic and back to political dynamic, as seen in past wars, is no longer discernable, and continued strategic adaptation will mean the use of more forward thinking and long-term oriented strategies that effectively combine economic, diplomatic and military methods.[91] Ideally, the foundations of which should have been pre-shaped years in advance of potential trouble

To further support the thesis and ideas presented thus far, it will be helpful to explore one final thought on war and developing doctrine and strategy. Traditionally the study of war looks at and attends to the "conduct" of war—an examination of tactics, doctrine, strengths and lessons learned. Additionally historians and scholars focus on the political dynamic and examine the cause of the conflict. In almost all cases the final results usually focus on the spectrum of simple to complex mistakes, actions and counteractions, and at least some focus on the "fog of war." Some have attributed success or failure in war to the distinction between greater and/or lesser powers or in essence the comparison of industrial-cultural development.

Scant attention has been given to the fact that a war is either short and sharp or prolonged, and protracted as a result of a *deliberate strategy.* For instance, success in the first gulf war, a short but sharp conflict, was attributed primarily to our technological advantage. On the other hand, Vietnam is viewed as a political failure. In light of this, if one applies the theories of Mao to Post-Modern war, and examines them in light of the current war and counterinsurgency effort in Iraq, our perspective on the war on terrorism might change. For example, Sun Tzu writing *The Art of War* over 2000 years ago noted that there is no instance of a country having benefited from a prolonged war.[92] Mao Tse Tung took that concept and used it to great purpose, and is the best-known proponent of the concept of a protracted war. He recognized that

civil war had distinct phases; first was the use of
d the second was regular war. He believed that
alue in continuing to pursue a guerrilla strategy
ld be immediately achieved with open warfare.
l war provides the basis and foundation for this strategy as it lapsed into a national war and then turned back to a civil war with bouts of short, sharp regular war in between. Mao also proposes that the theory of a quick victory is wrong, and as strength and weaknesses waxes and wanes between the warring factions, these subsequent stages also require a different form of fighting. Specifically he cites the decline of morale, tactics, ability and optimism as he applies different forms of fighting at each stage. Of fundamental importance is the support and perception of and from the international community. Additionally, your aims are won through the slow and progressive accumulation of advantages rather than the short, sharp finale of Western war that still focuses on centers of gravity and the single knockout blow. Political advantage and aims are achieved by protracting the conflict and draining political, financial and moral assets; there is innumerable value in waging innumerable small, but decisive battles. Lastly, Mao has deftly fused his own ideas with those of Clausewitz when he propounds that "when politics have developed to a certain stage beyond which it cannot proceed by the usual means, war breaks out to sweep the impediments away." He also means insurgency, terrorism, conflict and civil war to be used interchangeably.[93]

Another supporting example to help illuminate this theory can be found by examining the Arab-Israeli Conflict. From the period of the Balfour declaration to 1948 one could characterize that thirty-year period of one of guerrilla war, or Phase I. From 1948 to present, the conflict has resulted in open warfare in 1956, 1967, and 1973 with the intervening years characterized by guerrilla war and an increasing escalation of international intervention, politics and support. For instance, it is widely known that during the Cold

War era, the Russians and Americans were more than willing to allow the conflict to continue unabated and provide support as it suited perceived political and strategic interests at the time. America and Russia supported each others' proxy as a means to counter each other; France withheld critical support from the Israelis in '67 in a last minute change in foreign policy; and the Germans, Chinese and several others provided material and economic support to whomever would do business with them. The wild cards in this scenario were the miscalculations of Arab unity and nationalism, European and Asian countries that undermined the calculations by supporting the proxy of there choice instead of aligning with one of the super powers, and the fact that the Israelis were not easily reigned in and launched their own policies and actions (such as the Gaza, Lebanon, the Golan Heights and sundry preemptive strikes) to the detriment of everyone except themselves. Lastly the political and strategic ante was increased in '73 when the unforeseen and/or ignored consequences of international politics collided with simmering Arab Pan Nationalism.

In conclusion, the technology, tenets and principles are not wrong. Asymmetrical warfare does not change the basic doctrine or principles of war, but rather changes the order and priority of what elements and principles take precedence in shaping and planning for conflict. Asymmetrical war shifts the dynamic from the large unit to the small unit. Additionally, political considerations, media and local/regional attitudes will dictate military planning strategy. Terrorism, insurgency, guerrilla war and asymmetrical war are merely techniques and forms of war that still utilize the basic tenets, albeit in different forms that do not conform to conventional norms and mores. A fundamental weakness is in our inability to formulate policy and actions with a twenty to fifty-year outlook. Our increased impatience and benchmarking others' standards to our own culture has created unrealistic timelines and expectations for success. We pride ourselves in adaptability and ad-hoc decision-making, yet it is precisely this method of problem solving that fails to understand the long-term impact or consider potential unintended consequences.

Another major factor includes making the paradigm shift from linear to holistic planning and execution with long-range intent. To illustrate this, consider how major geo-political strategy relies extensively on Clausewitz and the principles of war, with the primary and most often quoted principle that war is an extension of policy by other means. Senior military planners and future administrations no longer have the luxury of thinking in terms of in-line incremental extensions that run the gamut from diplomacy, dollar diplomacy and sanctions all the while signaling potential military intervention with build ups and force projection. Future administrations and military planners and planning must now consider how they will eliminate these imaginary boundaries and steps in order to shape, prepare and intervene in a seamless fashion. Ideally, the shaping and preparing is occurring years in advance, simultaneous and in conjunction with policy and diplomacy. Although some would argue that this smacks intensely of Prussian militarism, a cursory look at the types of missions and projected tasks that the military now performs in conjunction with the Department of Homeland Security would indicate that we may have already surpassed that threshold.

Western root cause analysis and subsequent strategy formulations will require more than just defining and creating superficial solutions. In essence we always seem to create the solution before we really understand the problem and then we demand immediate results. Future success will demand great and fundamental paradigm shifts. At the risk of sounding overly militaristic, future success will need to have a coordinated military-industrial-political assessment and decision-making process that is futuristic oriented. Lastly, the military should embark on an intern program and/or implement sabbaticals in which military officers work in the business sector and learn how to translate lessons in globalization to military affairs. If it is safe to assume that military strategy has always consisted of the idea of "capabilities and places", the challenge for the future will be to further integrate and model long-term scenarios that consist of integrating the ideas of economic, political and social "capabilities and places."[94]

For all the hype on the "1–4-2–1 Defense Strategy" how will we effectively deal with coercion and regime change when we fail to consider the interconnected and interrelated effects of multiple "capabilities and places"? [95] Especially, with increasing Western apathy that is increasingly leading to isolationism and impatience for results. Whereas success was once measured in years and decades, success is now measured in months especially if we can expect increased military intervention in the future[96]

THE MARENGO CAMPAIGN:
ASYMMETRICAL ANALYSIS OF CONVENTIONAL OPERATIONS

Impetuosity and audacity often achieve what
ordinary means fail to achieve.

Niccolo Machiavelli: Discoursi, xliv,bk3, 1531

INTRODUCTION

The Napoleonic Era is rich with a singular number of trends that impacted the development and evolution of modern warfare, notably, those regarding tactics, weapons, professionalism, organization, and our concepts on the art and doctrine of war. In many respects, this was an era of "emerging doctrine"[97] and many of these lessons and concepts are still with us today in modern and updated form.

One recent aspect of current emerging doctrine is the concept of asymmetrical warfare. Current US Army Doctrine as outlined in Field Manual 3–0, Operations, has six short paragraphs on the doctrinal concept of asymmetry.[98] The definition in this keystone document on doctrine gives one the feeling that asymmetry is mostly applied to the realm of technological superiority. Arguably, using the proposed definition of Asymmetry in McNair paper 62, in reality, the primary differentiator of asymmetry is really "disproportionate effect" with a specific emphasis on the "psychological." [99] Arguably, commanders throughout history have always striven for this disproportionate effect, and have always looked for ways in which to bring immediate and decisive defeat. The Marengo Campaign and the Battle of Marengo achieved this disproportionate effect. We will explore the Marengo Campaign, for its relevance today to modern US Army Doctrine of Asymmetrical warfare and how those lessons can be useful today.

The War of the Second Coalition lasted a scant three years, from approximately 1798 to 1800. This period is generally considered as the transitional point between the War of the French Revolution and the Napoleonic Wars. The Marengo Campaign and the Battle of Marengo is the critical battle ending in strategic victory for the French, the war of the Second Coalition, and brings peace; albeit short and uneasy, something that Europe has not seen in ten years. However, we digress. In order to explore the relevant asymmetrical lessons and how they might be useful today, we will need to start at the beginning.

Paradise is under the shadow of our swords. Forward!
Caliph Omar Ibn Alkhattab, at the Battle of Kadisiya, 637

IN THE BEGINNING

The War of the Second Coalition essentially begins with Napoleon's effort to strike at Britain when he organizes the invasion of Egypt in 1798. His expedition begins on the 19th of May 1798 as his force sails secretly from Toulon. On the way his fleet is joined by convoys from Genoa, Corsica, Marseilles and Civita Vecchia.

This combined fleet now numbers about one hundred warships and four hundred transports carrying 36,000 troops and about 1230 horses—700 for the cavalry and the rest for staff and the artillery. Along the way he captures Malta. The expedition arrives at Alexandria on July 1, and Napoleon begins disembarking vicinity of Marabout. Napoleon leaves a rear guard to cover the landing site and then organizes the remaining 25,000-man expeditionary force in three columns. He captures Alexandria on the 2nd of July, sends one force about thirty-five miles to the east to capture Rosetta at the mouth of the Nile, and sends one force south towards Cairo. Cairo falls on 24 July. The campaign in Egypt is successful and Napoleon heads east towards Syria. Meanwhile, Nelson defeats the French fleet in Aboukir Bay on 1 and 2 August 1798, cuts Napoleon's lines of communication and strands the French in Egypt. Undeterred by this strategic setback and a couple of other operational events, Napoleon is delayed but continues with his invasion of Syria.[100] The French are mauled at Acre by both the Marmelukes and the plague, and Napoleon learns that things are going badly in France at the same time that he is bogged down in Syria (March-May 1799). Napoleon receives a recall message from the directory and he leaves 13,000 troops to the command of Kleber as he sails for France on 22 August, arriving about 1 Oct 1799.[101]

In summary, as we set the political and military stage for events to come, the Egyptian and Syrian campaign were strategically useless. They did not draw attention away from the critical center of gravity—Europe—and resulted in the unnecessary loss of some of Napoleon's best troops and equipment. Equally disastrous was the complete destruction of the French Mediterranean fleet in Aboukir Bay in which the English now had undisputed naval superiority of the Mediterranean Sea, and a disaster that the French never recovered from.

During the period that Napoleon was on campaign, simultaneous events were leading to the critical battle that would take place at Marengo.

The Coalition was enraged at French aggression and began to organize an effort to restore the "old order." The Second Coalition is formed under English insistence consisting of England, Russia, Naples, Austria, the Vatican, Portugal and the Ottoman Empire aligned against France and her allies: Holland, Spain and Switzerland. During the years 1798 and 1799, with Napoleon on the Egyptian Campaign, the leadership of France falls to Lazare Carnot. Carnot decides to initiate three simultaneous offensive operations against: 1. Austria and Russia in Italy; 2. Austria and Russia in Germany; 3. England in the Netherlands.[102] Initial French success in Italy later results in defeat. The Austrians win at Magnano and a Coalition Russo-Austrian Army routs the French at Cassano. Turin and Milan are lost, the French are routed at Trebbia, and French leadership is either replaced or killed in action. However, in Switzerland and Holland, the French dominate and the Russians withdraw from the coalition blaming insufficient support from the Austrians. During this same period it was the Austrians that drained themselves of critical sustenance, medical, engineer, artillery, line officer and staff officer support in order to shore up Russian units that lacked of practically everything.

When Napoleon arrived in France on the 9th of October 1799, France was near collapse; invasion by the second coalition appeared to be imminent, the country was near bankruptcy, the troops had not been paid or fed in months and had been living off the land, and the Army was more than outright contemptuous of the government. The coup d' etat (coup d' etat de Brumaire) takes place on 9 and 10 November and Bonaparte is swept into power. Under the cover of diplomacy and peace overtures, Napoleon begins rebuilding the Republic and the Army.

The new century dawns and France is in a precarious political and military situation. French efforts of peace and diplomacy have failed. France had less than enthusiastic alliance partners in Spain, Holland and Switzerland. Napoleon is still searching for ways to provision his Army and fix the neglect of the past years. Internal revolt still needs to be subdued.[103] The coalition forces along the Rhine pose a direct and immediate threat to France and Napoleon needed a victory to boost and consolidate his power. Napoleon initiates a second Italian Campaign in 1800.

Alliances, to be sure, are good, but forces of one's own are still better. Frederick William of Brandenburg (The Great Elector) Political Testament, 1667

PREPARATIONS: PAINTING THE STRATEGIC PICTURE

Bonaparte and France have had since 1792 to organize, reform and experiment with organization, leadership, strategy and tactics. French morale is high, although logistics and support still need to be significantly improved. Desertion is still high in the French Army. Russia has withdrawn from the coalition, and the remaining members are still determined and optimistic about restoring the monarchy, defeating the revolution, and stopping French expansionism/hegemony. Although the coalition forces have a healthy respect for French ability, they cannot seem to overcome and adapt to French strategy and tactics. The Archduke Charles is probably Bonaparte's only real opponent; however, he is transferred to the Netherlands in 1799 and will not impact the operations. Various accounts have the archduke refusing command, sent to Holland or "exiled" to Bohemia because he had the audacity to suggest to the Aulic Council that Switzerland was the strategic key, and that the coalition should strongly consider Napoleons peace overtures. In any event, this would buy the coalition "political space" in order to reconstitute coalition forces. Also, as noted before, the

Russians take a beating at the hand of the French in Holland and in Switzerland, and Tsar Paul withdraws from the coalition.

The French Corp is now an official and formal part of the Army Organization in 1800, and Army combined arms operations might arguably be, officially born.[104]

In the planning phase, Archduke Charles recognized the importance of Switzerland. However, Vienna felt that the Hapsburgs political interest was in Italy and Germany, and not Switzerland. Switzerland is strategic and key terrain for the lines of communication that it offers the French and its direct routes to Vienna through the soft underbelly of the Hapsburg Empire. In reciprocal, these same lines of communication are important to the Hapsburgs for any planned attack on Paris, and the Archduke and Napoleon recognized this. Vienna felt that political decisions should override military decisions. In principle they were correct. This is certainly in consonance with Clausewitz's later and famous dictum that the use and application of force is an extension of political aims;[105] however, the Austrians fail to recognize the strategic importance of the interior lines of communication.

Napoleonic aims in the period 1799 to 1800 were more complex. Napoleon arrives in Paris on the eve of potential insurrection. Political subterfuge and deception wreak havoc as royalists, republicans and reformists posture for power. Napoleon becomes the principle in a coup d' etat and nearly loses control the day after they seize power. As Napoleon consolidates his power, his initial aims are to restore peace and prosperity, yet strangely, his motivating speech to the public demands that war is the road to peace. As the Napoleonic ambassadors parley with the allied nations, it becomes apparent that the demands of the First Counsel are patently unacceptable. France realized coalition intentions and was disturbed by intelligence reports and troop movements. While the coalition was maneuvering its forces into place, Bonaparte too, is concerned with his intelligence reports and develops a plan to preempt them and attack first. Principle French objectives are the decisive defeat of the allied armies and the occupation of Vienna. In reciprocal, the second coalition aims for the destruction of the French in Northern Italy, and simultaneous, concentrated attacks by the coalition forces in the other theaters and with a march on Paris. Accordingly, it may be said that the plan was to hold Moreau by the nose along the Rhine: decisively defeat the army of Italy; link up with the British and Abercrombies Corps; and march on Paris via Toulon. Secondary results would be the assistance from Royalists as the Austrians would march on Paris (note the expectation of exploiting partisans and guerrillas).

In January 1800, Napoleon secretly orders Berthier, his Minister of War to organize a Reserve Army in the vicinity of Dijon. The troops used to create this reserve arrive piecemeal from around the country. Because the army was living off the land it was required to billet the troops around a wide area in order to reduce the burden of supply that fell mainly on the civilians. The Army was broke and could not sustain or billet its own troops. Because the actual troops that arrived in Dijon were ragged and undisciplined conscripts, coalition HQs dismissed the reports of troop arrivals and assembly as disinformation.[106] Simultaneously, while Bonaparte is attempting to coordinate all French armies in the forthcoming campaign, his power still has to be consolidated, and he is tactfully jousting with a powerful yet insubordinate General Moreau. Bonaparte makes several recommendations on the strategic planning of the overall campaign on the Rhine and in Northern Italy. Moreau objects to everything. Napoleon realizes that Moreau is too powerful to relieve him of command as his power has yet to be consolidated.[107] Napoleon finally issues a general order directing that the campaign would begin in mid April. Moreau (reluctantly) would detach a Corp from the Army of the Rhine in order to defend Switzerland and occupy critical Alpine passes.[108] This Corps would advance to turn Kray's forces and cut the lines of communication between the two Coalition Armies; one on the Rhine, the other, Melas, in Northern Italy. Once this had been successfully developed, Bonaparte would advance through the Gothard and Simplon passes and link up with Massena vicinity of Genoa. Moreau was still an obstacle to the operation. He continued to hedge against supporting Napoleon. It was clear that he would neither act decisively, nor cooperate [109] before Napoleon is ready to strike, surprisingly however, the second coalition gels and the Austrians pre-empt Napoleon in Northern Italy.

Arm me, Audacity, from head to foot.
Shakespeare, *Cymbeline*, I 6. 1609

SPEED, SURPRISE, LE' AUDACE!

Bonaparte receives the news that Melas and the Coalition forces had attacked on April 6, and by the 24 had split his command in two and driven Massena back into Genoa and towards Nice.

The French Army of the Reserve lacks everything and Berthier and Bonaparte's aids perform small miracles in logistics. The bad news from Italy rushes the French and causes them to change plans. Bonaparte approves Berthier's recommendation of using the St. Bernard Pass and begins moving the Army of the Reserve towards Geneva. Learning that Massena was cornered in Genoa, and Suchet is west of the Var at Nice, Bonaparte sends messages to Massena requesting that he hold firm until 30 May and Suchet hold out until 4 June. Bonaparte knows that his plan is risky, but speed and surprise are of the essence. [110]

As the Army of the Reserve shifts south it collects reinforcements along the way. Napoleon Arrives in Geneva on the 9th of May, but a lack of supplies requires him to delay the march columns until the13th. Napoleon uses the intervening time to conduct reconnaissance and organize his troops. In the meantime, during the period of April 25 and May 8, Moreau had attacked Kray across the Rhine and beaten back Coalition forces. Napoleon's Advance Guard and Main Body enters the St. Bernard Pass on the 14th of May. Sometime vicinity of 21 May Chambron enters the Little St. Bernard Pass, Turreau enters the Mt. Cenis pass, Bethencourt enters the Simplon, and, Moncey, detached from Moreau's Army of the Rhine, enters the Gothard. Effectively Napoleon has five columns converging on Northern Italy along a front that is approximately 115 miles long. Two main columns, Chabran and the main body, would merge at Aosta, and the remaining three would confuse the coalition forces of Napoleon's real intent. The destruction of the coalition forces is the primary objective, and relieving Massena at Genoa was now secondary.[111]

Between 17 and 24 May, the French clear small pockets of resistance in the passes. Ft. Bard is the only real danger and coalition forces hold out at the fort until 1 June. This causes the French some concern as they attempt to smuggle cannon and troops past the fort at night. By May 24 Melas is confused as to the strategic situation. Thinking that the French will take the most direct route to relive Massena, Melas judges Turreau's force in the Mt. Cenis Pass to be the French Advance Guard. Napoleon orders his main body to converge on Ivrea.

Between 27 May and 5 June, Melas concentrates his troops at Turin; Suchet chases Elsnitz forty miles east back to vicinity Savona and Ceva (these two cities are about thirty miles west of Genoa). Massena surrenders Genoa to Ott, receives favorable terms, and Napoleons columns converge.[112] His columns drive east and south ravaging the Coalition rear area and occupy an area vicinity from Ivrea, east to Milan, south, to just short of Piacenza, and a Corp occupies the north side of the intersection of the Po and the Ticino Rivers just north of Stradella.

During the period 5 to 12 June, Napoleon consolidates and begins moving west. Desaix arrives from Egypt and Napoleon promptly reorganizes his Army in order to create a Corp for Desaix; With Victors Corp as Advance Guard and Lannes and Desaix following, the French reach Tortuna without any resistance. By this point, both sides were increasingly worried because no contact had been made, and, both Napoleon and Melas had underestimated strength, march speed and intelligence estimates. Simultaneously, Melas and the Coalition concentrate at Alessandria and small pockets of Austrians occupy Genoa, Savona, Ceva, Coni, Turin and Casale

Therefore, determine the enemy's plans and you will know which strategy will be successful and which will not.

Sun Tzu, *The Art of War*, vi

MARENGO

The situation on the 13th of June was anxious; both sides had avoided contact. Melas' hasty retreat off the Marengo plain reinforced Napoleons belief that Melas would not attack. When night fell on the 13th, Napoleon's units occupied Marengo and a general line along the Fontanone Creek from Castel Ceriolo to vicinity a position about a half mile south of Stortigliona; and a French Division straddled the main road about three-fourths of a mile east of Marengo. As the French went to sleep in the mud and rain that night they occupied a poor position and their local security was worse. On the morning of the 14th one of Murat's aides warned Napoleon that the Austrians were advancing. Bonaparte dismissed this as a feint. Additionally one of Desaix's aides reported that their reconnaissance effort had not found any Austrians, and Bonaparte refused to believe this as well. Consequently, he ordered Desaix to screen a line east and south of Marengo along a line of Torre di Garfoli, Rivalta and Serravile. About 9:00 a.m. on the 14th of June things were heating up; the Austrians gently started the attack about three or four in the morning. While Desaix is marching to his screen line, Austrian lead units had begun to attack the French division straddled the road. The Austrians must attempt to deploy from the march, but terrain prevents this; additionally, the French are slow to react to the attack. [113] By 10: 00 a.m. the French have been pushed into Marengo, the Austrians are still struggling to deploy all their troops into a line along the creek, and Bonaparte, still convinced that this is a feint, finally decides that he should investigate the increasing roar of the cannon. As Napoleon mounts up, he rides to a position vicinity of Casa Buzana and glimpses a full view of the battlefield from this long, wide, gentle knoll and realizes the horror of his mistake. Bonaparte immediately sends messengers to recall Desaix and Lapoype and calls forward Monnier and the Consular Guard.

The Battle for Marengo rages; between 1200 and 1400, the Austrians make four assaults, all repulsed by the French. O'Reilly's Austrian Cavalry rides south and west along the Bormida, and attacks and contain a weak French detachment at Cassina Bianca. O'Reilly's cavalry continues south and is no longer part of the battle. It is assumed that he pursues the remnants of the French Cavalry pushed out of Cassina Bianca. Meanwhile, Ott followed the low road out of Alessandria, attacks into Castel Ceriolo and promptly reports no French. Ott immediately moves to envelop the French right flank. By now, just about 1400, Marengo falls to the fifth Austrian assault and the French right flank is about to be enveloped. The only thing standing in the way of an Austrian Victory is Champeaux's ravaged remnants of a Cavalry Brigade and the demi-brigades that Monnier deployed piecemeal vicinity Castel Ceriolo and Villanova. Monnier would pay dearly for deploying his units in piecemeal fashion, rather than as a complete Division.[114] Lannes, who had been holding the middle, was now in a precarious position. Both his flanks were exposed and he had to withdrawal. At Villanova, the Consular Guard takes a beating and can no longer tolerate the heavy cannon and infantry fire decimating the square; they retire back towards San Guiliano. By Now, the French were deployed along a line just west of San Guiliano along the Sale and Novi Road. The French had no artillery, had managed to barely cover their withdrawal by successive rearward bounds, and what little remained of the cavalry was required to maneuver to cover the flanks. Bonaparte's few remaining escorts were all that remained of the French Reserve. This was clearly the time for Austrian Cavalry to drive home the pursuit and to rout the French. Problem was, the Austrians didn't have any Cavalry left. The French Cavalry, Kellerman's Brigade, had utterly decimated one Austrian Brigade when they attempted to cross the creek at Marengo; two other Austrian units had been pounded hard vicinity of Castel Ceriolo, and the remaining Austrian Cavalry was intermixed with the Infantry and would be impossible to recall. Melas decides that victory has fallen to the Austrians and does not need his cavalry; the infantry can finish the job. Melas passes command to Kaim. Melas is battered from the last several days of hard fighting. He is seventy-one years old, and is battered and bruised from long days in the saddle and from having two horses shot out from underneath him. It is about 1500 and Melas trots back to Alessandria. In the resulting confusion and chaos, Kaim does not give explicit instructions for the organization of the pursuit.[115] For an hour and a half the Austrians seem to leisurely begin their move out of Marengo, and continue east down the high road, but they do so in column, not attack formations. This breaking of contact allows the French a little breather and brings order back to the French units.

About this time, Desaix arrives with a Division.[116] They use this division to straddle the high road in Demi Brigades and deploy another brigade, the 9th light, to the left off the road. Behind and between the 9th light and the units to the right were Kellerman's Cavalry and the survivors from Champeaux and Duvignau's

Cavalry brigades—or about six hundred and fifty horses. Desaix requests artillery and Marmont began scraping up what he could find. The dispositions were completed about 1700. The Austrian advance guard reaches French positions and is greeted with a volley fire from the 9th light and Marmont's few cannon. Accounts differ and it appears that a few cannon translates to four or six twelve-pounders. The Austrian advance guard breaks and runs, but the Austrian Grenadiers shake it off and continue to attack. The Austrians quickly unlimber two batteries and pour cannon fire and musket fire into the 9th light. The 9th is pounded hard and is forced to pull back. Desaix sees the crisis and commands the Division forward on the attack, Boudet, the division Commander, is almost immediately shot dead. As the Austrian fire quickens, the 9th reels in disorder, and the rest of the division begins to follow. The Austrians quicken the pace, and fix bayonets as they march through the smoke, about to make the victory complete. Marmont was trying to limber up the cannon for the command to charge. He now barely had time to get three cannon back into position and he gave the Austrians a whiff of canister. At this very moment, Kellerman brings the remnants of the cavalry forward, six hundred and fifty strong, and rides down the Austrians while they attempt to collect themselves. The Austrian Dragoons on the advance guard's left flank failed to act on their opportunity. Kellerman quickly regroups and rides them down as well. Boudet's leaderless division and the few survivors of the 9th light regroup and charge again; Bonaparte releases Bessieres Guard Cavalry, and the Austrian victory instantly turns to a rout. Austrian Dragoons and follow-on cavalry caught in the rout are carried away with the dragoons, and they ride down their own units as they attempt to flee the field The French had snatched victory from the jaws of defeat. By 2100, the French had regained the field back to Marengo and the Austrians retained Alessandria and the Bormida Bridgehead.

An enemy that commits a false step... is ruined, and it comes on him with an impetuosity that allows him no time to recover.

Cuthbert Collingwood, 1748–1810 (writing after Copehagen of Nelson's quickness to profit by enemy blunders)

THE MORNING AFTER

Bonaparte begins reorganizing the exhausted remnants—no ammunition, few available troops for a reserve and barely any Artillery.

The Austrians call a council of war. Though they still had the fortress troops and fresh reinforcements, they had lost more 9,400 reported casualties, including many Corp and Division Commanders, and the Chief of Staff had been taken prisoner by Kellerman's brilliant charge.[117] The Austrians voted to negotiate. They were in shock and utter disbelief.

Bonaparte was not in a bartering mood, they had a 48-hour armistice. Bonaparte would talk if the Austrians surrendered the bridgehead over the Bormida.

The armistice turned into the convention of Alessandria and the Austrians pulled back behind the Mincio River. Diplomacy, skirmishes and negotiations reigned from June until November of 1800. Napoleon finally saw that the Austrians were buying time and he denounced the armistice. The winter campaign ended in February of 1801 with the French victorious and the Peace of Luneville.

THE RELEVANCE TO ASYMMETRY

Introduction: Technology increases "friction" and requires additional "genius" for war. Modern battle space is a multi-dimensional concept, a concept that requires spatial, creative and analytical skills. This is the essence of "Coup d' Oeil"—the ability to synthesize in 4D, if you will. It is said that Napoleon had this uncanny ability. This is the anti-thesis to the linear type of thinking that was, and is, characteristic of the past. Regrettably, with the modern emphasis on technology and technological solutions, we still think in linear terms. The reliance on technology does not cultivate the instinct, intuition and/or "Coup d' Oeil" that will give a soldier the leading edge in times of chaos and crisis, or in decision making when technology is not available or has been disrupted. Every age has attempted to find the asymmetrical silver bullet that would practically guarantee success. For instance, In the Napoleonic Era, it was Bonaparte's doctrine of "*La manoeuvre sur les derrieres.*" And a scant sixty years later, with Moltke and Schleiffen, it would be the doctrine of "*Kesselschlacht.*"

Some would argue that the relevance to asymmetry in the Battle of Marengo is only appropriate for conventional war. On the contrary, it provides clear lessons of relevance today in improvisation, complacency, friction, coalition warfare and leadership; all elements that are critical regardless of the type of operation—conventional, unconventional or asymmetrical.

Strategic lessons in Asymmetry: Bonaparte concentrated on defeating his opponents before they had a chance to consolidate their forces. In this campaign Bonaparte did not know where his enemy really was, but, marched to the sound of the guns, and had an appreciation of political and militarily strategic terrain and outcomes. More on this later, but, luck, chance, and a few good French commanders insured that Napoleon's victory was decisive. The victory turned to rout provided the psychological disproportionate effect that is the hallmark of asymmetry. The second coalition lacked a common strategy and war aims on a strategic level. This particular point is driven home hard when one examines the current debate within the allies, supporters and members of the UN Security Council as it regards the current US intent and methods of diplomacy in disarming Iraq.

Hapsburg commanders had little latitude in planning and execution and Great Britain lacked an effective land force. Austria has the lion's share of the responsibility and role in the second coalition, and was drained of resources by its Russian Allies. This last point is particularly similar and relevant when one looks at all the current theaters of operation that the US is currently involved in on the war on terror. Additionally, other internal government agencies will be bled of competing resources as they try and support both internal and external requirements.

Politically, restoration of the monarchy, the overthrow of the revolution, and a fear of France and Napoleon is the common thread in the alliance, and the coalition collapses at the end of 1799 with the withdrawal of the Russians. The French campaign to Egypt was utterly useless and served absolutely no real purpose. The French lose some of their best units and equipment, and the French Mediterranean fleet is annihilated leaving the English in command of the Mediterranean Sea. Clearly what is important here is the requirement to have a vision of what is to be accomplished: clear military and political goals and expectations. Like the current war on terror, goals and objectives must be constantly assessed and updated to insure continuity and integration. Additionally, it also shows how fragile the politics of developing, forming, and sustaining a coalition and its partners can be, especially when there are significant ideological differences between parties, and these parties view American techniques and policies as aggressive.

The point here is the common thread of fear of American hegemony. This will most certainly be the most important lesson in asymmetry; especially, as the world comes to rely on a greater extent on collective security and organizations such as the UN. Of course the real test will be to insure that these organizations also have the mettle and resources to back up what they demand. Without digressing, it appears that the recent UN support for the disarmament of Iraq is the first time that the UN Security Council is more than a paper tiger, albeit at the insistence and demand of US unilateral action. Resistance and recalcitrance in the rest of the world nations shows a great disparity in policies, thinking and ideology. It would appear as though the rest of the nations are testing our resolve. And as more powerful countries such as France, Russia, Germany and China are more vocal and resistant of American ways and means, will now, and has now, provided the courage and catalyst for lesser countries to buck and resist US efforts at coalition building.

Operational lessons in Asymmetry: Coalition commanders displayed little imagination or ability with the exception of Charles. Napoleon literally was flying by the seat of his pants; he had no real intelligence, poor dispositions, and his units failed to secure the Bormida Bridge crossing when they first arrived at Marengo. Bonaparte has no real choice except to trade space for time and fall back. This appears to be clearly a defeat for the French; Desaix arrives in the late afternoon, tells Bonaparte that there is still time for a victory and quickly maneuvers to the attack. The coalition is caught off guard and in march formation. Desaix saves the day when the coalition is driven from the field and a rout turns into an armistice. Operational and Strategic disproportionate effect had been achieved at by actions at the tactical level. Although the Austrians were not actually decisively annihilated, they had been decisively beaten psychologically. Quick thinking on the part of French commanders, coupled with surprise, audacity, courage and endurance had not just saved the day, it created political and military conditions that ultimately lead to the end of the second coalition. In a comparison with today's sense of timing one might find it hard to imagine that the Marengo Campaign and the War of the Second Coalition was actually conducted with lightening speed. As a comparison, the French calculated Infantry march rates at approximately twelve miles a day. Additionally, according to David Chandler (*The Campaigns of Napoleon*, 277), Bonaparte rides from Paris to Geneva during the period 5–8 May, a distance of over two hundred and forty miles, or approximately sixty miles a day on horseback. That is incredible riding. (I should know, I am an equestrian rider and have ridden long distances.) By comparison, during Desert Storm, some units raced along at forty miles per hour and made one hundred miles a day. Restraints on how fast and far units traveled was largely determined by the operational plan, risk assessment, the fear of outrunning logistics and other force protection considerations[118]

Modern technology has skewed and utterly compressed our sense of time, distance and space. Yet, according to Donald Rumsfeld and the Central Command Commander, the First Cavalry charge of the 21st Century was conducted by Special Forces troops on horses. A fitting irony for a war that is being conducted with the most technologically advanced soldiers in the world. Future and present asymmetrical warfare will require important and incredible changes in American thought patterns, and will certainly require a different sense of space and time. In an operational sense, the key is decentralization. The Al-Qaeda are using modern technology and "swarming" techniques in order to quickly group and disband cells. American operational success will come at the small unit level. Traditionally, in the past, this has been considered the Battalion. Small unit actions now comprise "A" teams and the squad and platoons of the "Company Team."

Principles of War and Asymmetry: Napoleon's plan utilizes the principles and tenets of objective, speed, surprise and audacity; his unexpected arrival in the rear area or *manoeuvre sur les derrieres* was designed to bring maximum effect. He focuses his concentration on the decisive defeat of forces in Northern Italy, utilizing Moreau and Massena in both the strategic defense-tactical offense, and the strategic defense-tactical defense.[119] Although Napoleon had not destroyed the Austrian Army, he achieved disproportionate effect, and undoubtedly learned some lessons in generalship that he would never forget. Namely, "when you contemplate giving battle, it is a general rule to collect all your strength and to leave none unemployed. One Battalion sometimes decides the issue of the day." [120] I would imagine that by about 1400 hours on the 14th he regretted not listening to the reports that morning, and sending Desaix and Lapoype off on screening and reconnaissance missions.[121]

Friction: surprise, audacity, shock, speed, leadership and quick thinking by some French commanders compensate for operational failures and become combat multipliers, the unanticipated and timely arrival of Desaix's Corp, coupled with Kellermans quick thinking turns defeat into victory for Bonaparte. This type of lesson becomes increasingly crucial at the small unit level. Although operations will be planned and commanded by theater commanders, strategic, operational and asymmetrical success is directly dependent on the ability of the junior leader. Especially crucial is the young leader's ability to quickly move from a kinetic to a non-kinetic response and posture, understanding that the choice of a hard-knock approach versus a soft knock approach will have both immediate tactical consequences as well as long-term strategic consequences.

You may pardon much to others, nothing to yourself. Ausonius, fl, 4th Century a.d., *Epigrams*

CONCLUSION

In conclusion, Napoleon's victory at Marengo was a result of luck and leadership. Disproportionate effect and psychological impact, albeit, by both chance and quick thinking, changed the political face of Europe. Although warfare during the period was conducted primarily in a tactically linear fashion, there were incremental improvements in organization, tactics and technology. The main aim was to seek the decisive battle that resulted in complete destruction and rout of the enemy force. During the course of the wars of the Revolution, Napoleon had begun to earn his reputation as a commander. Napoleon rejuvenated warfare with his infamous Corps Organization and tactical *manoeuvre sur les derrieres.* He was able to achieve disproportionate effect by wrecking havoc in the rear area and defeating his opponent with psychological effect before he engaged in the decisive battle. Later, Bonaparte's success at Austerlitz, also an "asymmetrical" victory, would essentially effect strategic and tactical thought and discussion through the American Civil War and onto the First World War.

Napoleon consistently confused his enemies with his constantly changing tactics. His enemies could not predict any consistency in his *modus operandi* and they never knew what to expect. He was consistently outnumbered by his enemies, yet he was able to defeat them decisively and with disproportionate effect. Napoleons methods were essentially emerging doctrine.[122]

Technology or emerging technology played no role. Success was a result of leadership and intangible skills. Increasingly, the small unit leader will need to have honed his instincts in order to quickly move between skill sets, appropriate response, and the intangible *combat multiplier le' Audace, le' audace, encore le'audace.*

A TALE OF THREE CITIES:

STALINGRAD, HUE AND MOGADISHU LESSONS IN URBAN OPERATIONS AND ASYMMETRICAL WARFARE DOCTRINE

STALINGRAD

In December of 1941 as Operation Barbarossa comes to an abrupt halt at the gates of Moscow, the German High Command plans and launches Operation Blau—the summer campaign and full-scale offensive on the southern flank of the Eastern front. This operation was supposed to be the "strategic decision-seeking offensive." [123]

Stalingrad is a prominent feature on the Don River bend that is now key to the strategic decision making process. This decision was supposed to put the Caucasus and the strategic oil fields in Hitler's possession, although Hitler and the German general staff never originally intended to actually fight in the city. Stalingrad has become a textbook example for strategic, operational and tactical lessons in urban combat.

With this in mind, it has been characterized that warfare in the 21st Century will see an increase in urban military operations.[124] Accordingly, the preface to Field Manual (FM) 3–06.11 (Draft), Combined Arms Operations in Urban Terrain, indicates that all operations that involve the Military will most likely involve urban operations for the foreseeable future.[125] Of parallel concern is the continued and increasing debate and discussion of asymmetrical warfare. Subsequently, urban operations will be an asymmetrical element to be exploited. Interestingly, the new draft manual has 1 paragraph in chapter 1 that addresses the symmetrical and asymmetrical threat, and FM 3–0, Operations, a keystone manual for US Army doctrine has six short paragraphs in chapter 4.

The definitions left to the reader in these manuals are both broad and short. Although these definitions leave the reader with the distinct impression that asymmetry is largely achieved through technological means, it can be safely argued that commanders at all levels of planning and execution, since the first recorded battles have sought to seek the decisive strategy, weapon and technological advancement that would bring swift victory from a massive blow, and that this blow would be rapid, decisive and result in an overwhelming and disproportionately successful effect. Another consideration of the available definitions leaves the reader with the clear impression that asymmetry is largely planned and achieved at the strategic and operational levels of command. Yet clearly, the squad leaders to the battalion commander are as capable of offering or inadvertently providing an asymmetrical wedge as well as the theater staff or commander in chief.

The summer offensive, the battle for Stalingrad, and post-Stalingrad operations offer important lessons in asymmetrical warfare that have parallels and usefulness for today's modern, joint and coalition army. These lessons correlate strongly with the doctrinal views expressed in a number of current manuals and documents such as FM 3–0, FM 3–06; USJFCOM J9's coordinating Draft of Rapid Decisive Operations; JV 2020, and the NSS Phase One document.[126] This chapter will explore some of the relevant historical lessons of Stalingrad and how they apply today to current doctrine.

Stalingrad and its relevance to asymmetrical warfare offers lessons across the entire operational framework of strategic, operational and tactical elements of planning and execution. In particular, the lessons of Stalingrad are applicable to modern combined arms, joint force and coalition operations.

As Hitler meddled in the planning and execution phase of the operations he was especially worrisome and indecisive on the Eastern front. As the Russian troops increased in numbers and largely became cannon fodder for the Wehrmacht, pockets of stiff Russian resistance and sheer mass began to take their toll on the German Army. Additionally, Hitler had serious misgivings over his general staff's ability to carry out his exact instructions. He attempted to unequivocally subjugate his officers, and subsequently hired and fired several general officers, redrew army boundaries, and reallocated units and missions to suit his personal taste, increasing paranoia and the subsequently increasing fixation with the name and city of Stalingrad.

It is easy to see with perfect historical hindsight, how the Germans failed in a number of areas. For instance, judgment; this failure was born of many elements of leadership, impatience, impetuousness, players, personalities, events and ideology.

Arguably there were many errors in judgment; however, one of the early signs of errors in judgment occurs when Hitler continually assesses coalition forces on the Eastern front in purely mathematical terms. These units with rare exception were neither at full strength nor of the Wehrmacht quality that Hitler insisted they were. A deeper strategic problem was with the League of Nations

Army, and the corresponding difference of opinion between Hitler and the high command on the equipment, leadership and training of the coalition forces on the Eastern front. There were only a few exceptions to the foul characterizations attributed to the Hungarians, Romanians and Italians. The Soviets would exploit this disparity that would ultimately lead to the encirclement and destruction of the 6th Army.[127] When the Soviets launched Operation Uranus on 19 November 1942 they had massed the 5th Tank Army and 21st Army on the northwest, approximately one hundred miles in straight-line distance from Stalingrad, vicinity of Serafimovich on the Don, and attacked through the approximately nine Romanian divisions. On the south, they massed three armies, the 51st, 57th and 64th, approximately thirty miles southeast of Stalingrad on the west side of the Volga, and attacked through three Romanian and one German division. The 64th and the 62nd would hold the line at Stalingrad, and would continue to hold the Germans by the nose as they conducted frequent attacks in the city. Gleaning their intelligence of the situation from a number of ways, but primarily from patrols that were sent to deliberately capture "tongues," Soviet slang for prisoners, [128] they clearly knew that the Romanian divisions in these sectors were both ill-equipped and ill-led and despised by the Germans. Besides the fact that the Soviet plan was clearly a deep battle strategy that intended to break through far to the rear and cause immense panic in the depot and army rear areas, the Soviets primarily exploited the weakness of integrating units with in the Coalition. Modern US forces are, and will be, required to work extensively with a number of other coalition armies. Those that are primarily "non-Western" could potentially be the American Achilles heel in any operation, primarily for the difference in training and technology. Even the other so-called Western armies have language, cultural, technological, equipment, leadership and rules of engagement differences that can and will be exploited.

A different strategic asymmetry was created by the Germans to the gradual and subsequent advantage of the Soviets. For instance, a German tactical and strategic blunder was the blind reprisal and retaliation methods that they used. In essence, although much of the population thought of the Germans as rescuers, the Germans treated everyone the same. They were all considered combatants and/or partisans. Furthermore, as the German Army used Napoleonic foraging techniques, additional portions of the population came to despise and hate their so-called rescuers. From a strategic point of view, this stiffened resolve, determination and ideology. These elements would later be successfully exploited further by the Red Army as the intensity at Stalingrad increased, and as German disciplinary and collective punishment measures escalated in harshness.[129] In a comparison with the National Military Strategy of shape, respond and prepare now, which is translated into the operational planning fundamentals in chapter 4, FM 3–0, Operations, Fundamentals of Full Spectrum Operations, the German high command in expectation of a clear, decisive and rapid victory in the East, began to "shape and prepare" the battlefield with these reprisals. Regardless if this was actually intended or not, the unintended consequences of these actions, planned or unplanned, shaped the battlefield. As is now common knowledge, these harsh measures worked to the gradual increase in resistance and firmly entrenched Soviet ideology, and provided immense amounts of material for Soviet disinformation and propaganda, even converting those that devoutly hated the Stalin regime, yet preferred the Soviets to the Germans because of the brutality and indiscriminate harshness of German measures.[130]

Another strategic German failure was the constantly changing organizational structure and missions after Operation Blau. From about 23 August 1943 when the First Units of the 16th Pz Division viewed the Volga from the high ground at Rynok, until the final collapse on 31 January 1943, there was little consistency in task organization assignment structure. For instance, the two Engineer

Battalions were actually army assets. Engineers were task organized by Squad, Platoon or Company, but no mention was made of task organization in conjunction with mission and capabilities. Arguably, as units were decimated and not replenished, the 6th Army had no choice but to reconstitute ad hoc units. This might work in a pinch, but once again, differences in unit equipment, training and leadership would ultimately work against them. Time and equipment would be the two major factors that could potentially overcome this deficiency, both of which the 6th Army did not have. Coupled with this was the fact that the *Luftwaffe*, although providing close air support, was actually operating as an independent arm and assigned there own air missions and sorties. Ground commanders may have requested specific air missions, and there may have been some general officer discussion over planning and some promised support, but the *Luftwaffe* was independent of the ground commander and independently assigned air tasking orders and missions. In a take on the "split military psyche" the *Luftwaffe* had their own ideas on what should or should not be a target.[131] They generally considered ground support and tank buster operations demeaning and menial. For instance, it is incomprehensible why the German Air Force continued to bomb rubbled areas within Stalingrad instead of focusing on the deep battle and interdicting follow on Soviet forces, reserves, artillery, command and control and logistics.

Discounting the power plays, animosity, flawed strategy and operational planning by Hitler, the German general staff did a remarkable job of continuing with the tradition of the "operational art" at the Division level and below. Additionally, this speaks highly of the German soldiers, NCOs and officers that executed the tactical elements of these flawed operational plans, even as composite ad hoc units. If anything, that itself is a study of leadership, training, camaraderie and discipline. Additionally, when one considers that the actual street fight and building for building close combat was conducted by squads, sections and platoons, which in many cases were also ad hoc formations, that also speaks highly of the training and ability of the soldiers and NCOs that performed the lion's share of the leadership role in the urban fight.

As one studies the incremental and increasing changes in the operational planning and subsequent fixation with Stalingrad, there is a remarkable resemblance to the phases or stages of conflict escalation. As each belligerent draws a line in the sand, the stakes become increasingly higher and the amount of resources and energy committed is commensurate with the amount of hatred, enmity and desire to win at whatever the cost. Both sides were determined to win at Stalingrad. For the Germans, it was an underestimation of Soviet resolve, available manpower, and resources. Further, it was a matter of concentration of command authority and a delay in strategic decision making. For the Soviets, it was an act of desperation and survival. They had nothing left to lose but the country itself along with whatever dignity and honor they had left; harsh Soviet disciplinary tactics also helped stem the tide of "defeatism."

Although discussed as an individual element, this is actually part of a series of calculations, miscalculations, and ideology that were "shaped" over time. Besides the obvious fact that the Germans could and should have bypassed Stalingrad, and besides the fact that both sides made a deliberate and conscious strategic decision to bleed each other white on the Volga, neither side had the foresight to use these elements to shape the battlefield. That is at least not until after the battle of Stalingrad, but more on that later.

Tactically, Stalingrad was a model of improvisation, and necessity is the mother of invention. Improvisations during the Stalingrad campaign were as much a result of need and creativity as a means to overcome deficiencies in supply or doctrine. Each side learned to put wire mesh over windows to keep out hand grenades. The Soviets developed "hugging" techniques and kept the front trace within fifty yards of the German troops in order to reduce the effectiveness of air and artillery. Both sides used reconnaissance in force as a means of implementing economy of force. Both sides developed squad level "storm troop" tactics and organizations to better deal with the chaos of close combat in an urban setting, and to better exploit the existing capabilities of assigned weapons systems through a modified task organization.[132]

In all respects, it was the gathering and cumulative effect of a number of variables or friction, [133] which speaks largely to a strategic and operational control born of shape, respond, and prepare now, of which the German high command had both miscalculated and seriously misjudged. The implications of a very long-range plan that must begin well before the conflict stage surely implies the need for a full court press of all covert and PSYOP operations, and the conscious decision to implement these shaping strategies well before consideration of troop deployment and trigger pull, ideally, in the diplomatic stage. As an example post-Stalingrad, the formation of the German League of Officers was the Soviet exploitation of PSYOPs. This "league" had absolutely no credibility and no believable plausibility. If both the Soviets and Germans had expected to create an offensive employment of friction then this charade would have had to have been born long before Operation Ring. Timing is a key element in establishing this believable plausibility.[134] Post-Stalingrad was not the time for either side to attempt to manipulate world opinion with such nonsensical charades.

What is relevant to modern operations is this cumulative and culminating effect of many elements that created the psychological and disproportionate effect that led the Soviets and Germans on the road back to Berlin.[135]

There are several examples in Biblical history of asymmetry and shaping the battlefield. For instance in the Book of Joshua, chapter

2, Rahab tells Joshua's spies that they have a great fear of the Israelites, and recounts the crossing of the Red Sea and the destruction of several kings. The Book of Judges also provides some examples of missteps of incomplete conquest, apostasy, rebellion (by the Israelites), oppression, deliverance, asymmetry and then shaping operations through the marriage of Samson.[136] In 1 Kings, the Queen of Sheba visits Solomon to validate all that she had heard. Using another lesson in history to explore and explain asymmetry and the "shape and respond" strategy comes from the Book of Jonah. Although Biblical scholars would indicate that the Book of Jonah is an allegory of lessons, it does provide a good example of potential shaping. If we examine Jonah 3:1–5, 10 we will see that first Jonah is an unwilling participant. He is called to be an ambassador. The commander in chief sends Jonah to Nineveh. Jonah is to tell the Ninevites that unless they repent, God's punishment will be swift and severe. The Ninevites repent. Jonah as ambassador is the reluctant messenger. Jonah feels that the Ninevites should be destroyed and does not want to deliver the message because he knows that if they repent then God will show mercy.

Although the timeline is greatly expanded it will show how God or the commander in chief used a shaping strategy to achieve his ends. First, Nineveh falls about 612 BC. Although the Book of Jonah is written circa 400–200 BC, it is known that the historical elements of the allegory occurred between 750 BC and 612 BC. By now, there has been somewhat of a PSYOPs campaign. The Ninevites had surely known of Sodom and Gomorra, they had heard of the flood, the fall of Babel and the crossing of the Red Sea. Moving forward in time to circa 1200 BC we have the exodus and conquest, and now, about 750 BC we have the predictions of Deuteronomy—the fall of Jericho and the fall of the Northern kingdom.[137] So, with this series of events there has been a clear shaping of the battlefield. Of course there are a number of assumptions and elements that might provide "scenario variants." For instance, what if Jonah had not delivered the message with conviction? Surely our present-day diplomats' and leaders' personal opinions will account for an altering of the intended outcome. And of course there are a host of other questions and biblical timeline debates, which are recognized, yet do not provide the room for such an extended debate. Never the less, this was intended to show an "shaping" example and how the current doctrine of asymmetry—shape, prepare, and respond—will need to clearly be assessed, reevaluated and implemented over the course of many years.

Another example would be to consider current urban architecture. It is a known fact that a number of buildings and facilities built with in the last ten years are also built to inhibit or prevent the use of wireless phones and other wireless apparatus. So with this in mind, how will the military shape, prepare, and respond to that urban dilemma? Will staff's be specialized by region and/or city? Will commands and/or divisions become highly specialized as well? Will CENTCOM get the city and architecture drawings for every major city in its theater? Has acquisition of new military technology taken the potential wireless blackout areas into consideration? How will these weaknesses be compensated for in current and/or near future urban operations until technology, strategy or both catch up with the R&D and equipment issue phase?

Arguably, one might say that the United States would never deliberately make the kind of decision that the Germans made regarding a deliberate decision to a battle of absolute attrition and annihilation for a city, or that we would even tolerate the kind of casualties that are a result of urban combat. On the other hand, we also know that cities are increasingly important for their political, economic, b2b and ideological symbolism. With Internet and system networks, backups and system redundancy, as a minimum there will be the requirement to insert troops into the urban setting rapidly and decisively. Securing the multiple and redundant networks are as key and critical terrain as airfields, power stations, dam's, oil fields, radio stations and government buildings. Assuming that the shape, respond, and prepare now doctrine also provides for computer electronic warfare and electronic counter measures well in advance of any fight, It will still be important to physically secure these items to prevent destruction of data and help restore the infrastructure to normalcy during security and stability operations.

All in all there are number of lessons from the battle of Stalingrad that relate directly to an enemy that will seek to exploit asymmetry in his operations. Once again, although some might argue that this was a conventional war setting, the lessons are clear and unmistakable for modern army doctrine.

Strategically: Shaping the battlefield comes in many ways and designs. Since we are already practicing unrestricted warfare in the political and business world it will be important for the military to do the same. Shaping operations will need to address religious, ideological, humanitarian, business, political and social elements just to name a few. In order for the charade of the German League of Officers to work, that type of PYSOP would need to be planned and implemented very early in the operation. Not after the slugfest decided the outcome.

Coalitions: Will come in various shapes, sizes and resolve. Adding to the friction dilemma will be the leadership, training, equipment, politics and resolve of each unit from Battalion, Company and Squad level. Aggregating ability by organization, country or nationality will be a dangerous prospect. Because there is a growing concern in the world that the US is exercising its power in inappropriate ways, the "Yankee, Go home" attitude is increasing in intensity.[138] Consequently, this could potentially act as a contagion among the officers and soldiers within those units. If there is a hint of mistrust or lack of confidence in leadership, equipment, ability or government action it will take a Herculean effort to prevent the intolerant derision and mudslinging between units and forces, not to mention the monumental degradation in morale, unity and cohesion.

Tactically: With the understanding that a tactical failure can result in strategic and/or operational failure, small, very small units, that is, from squad to company will need to be highly educated and might strongly consider some kind of area specialization within selected teams. It is a known fact that there are not enough special operations soldiers to cover all missions across all theaters. Some conventional units other than the 82nd Airborne, Division 101st Air Assault Division, 10th Mountain Division and selected Marine Corps units will need to be trained for specialized regions and missions. In turn these Divisions or Corp-size units would work with TRADOC to act as the "proponency" for that task and region, and then use train the trainer techniques and reassignment selection to cross level knowledge and skills.[139]

Across the Strategic Spectrum: It is not sufficient anymore for soldiers to just know the commander's intent or know and understand the broad brush stroke of the strategic intent. They must be knowledgeable of regional and global events. They must be intimately knowledgeable of global events. They must understand how the tactical operations can enhance, mitigate or wreck the operational and strategic plan. Increasingly, soldiers will need to add the *Harvard Business Review*, *World Energy*, *The Washington Post*, *Forbes*, *Business Week* and *Baseline Magazine* to the book shelf and ruck sack of FM's and other materials that accompany them on training or deployment missions.

Unit organizations should become networks with the most expert and recent data, equipment and resources that are available for use. The networks should be intellectually and operationally redundant. We should consider experimental reorganization that would test the habitual op-con, assigned, attached methodology to one that is network capable. Meaning, as indicated earlier, that a company is a network, and each platoon is a network within a network, and each company is a network within a battalion network. A high degree of specialization is now required, knowing that tactical, operational and strategic specialization may be an asymmetrical weakness. For the modern soldier to be capable of full-spectrum dominance and capable of operations other than war, they will need to act like corporate business development and implementation teams that have all the expertise with in a small group of people. If they need staff support and additional resource commitments, they are empowered to go directly to the required resource; it is a flat organization that eliminates layers hierarchy. These teams have the authority, knowledge, expertise and resources to remove barriers, make decisions, and "flex the plan on the fly"(within the commanders overall intent) if that is what it takes. And these skills and skill sets *should not* be the proprietary right of special operations, but should also reside in so-called conventional units as well.

In conclusion, in order to exploit urban asymmetrical advantages and disadvantages, the US military will need to reconsider the paradigms and organizations that currently exist, especially in consideration of future urban warfare and operations. Technology is only one small aspect of asymmetry. What are clearly important are the intangible elements of friction that commanders shy away from. It should be clear by now that our squads and conventional units of all shapes and sizes need to have, and/or, be expert in a broad number of areas other than just the traditional military studies, and combat on urban terrain should be classified as a highly skilled and specialized field that deserves an extensive amount of attention.

HUE

The Tet offensive was a watershed mark in the history of the conflict in Vietnam. On January 30 and 31, 1968 "combat erupted throughout the entire country. Thirty-six of forty-four provincial capitals and sixty-four of 242 district towns were attacked, as well as five of South Vietnam's six autonomous cities, among them Hue and Saigon." [140]

The Tet offensive was North Vietnam's masterstroke of strategy. This was Hanoi's asymmetrical and indirect approach to a growing stalemate in the south and was a "general offensive" that relied primarily on the "social dimensions" of war.[141] The plan called for widespread and simultaneous military actions in both rural and urban areas. Hanoi attempted to achieve the general destruction and collapse of the South Vietnamese military and political infrastructure. They aimed to undermine and subvert the general population's confidence and morale by showing how vulnerable they were and how inadequate government protection was, and the final result would be an expected popular uprising throughout the entire country that would halt the fighting and force political concessions at the negotiating table.

Until this time, Hue had escaped the ravages of war. Hue was a symbolically rich city. It was the old imperial capitol; it was the intellectual center of Vietnam and had a tradition of Buddhist activism and anti-Americanism. It was also militarily significant because Highway 1 was a major line of communications and logistics. However, before we can proceed we will need to digress in order to set the stage for the analysis of the events and their importance to lessons in asymmetry and today's doctrine.

The US Army had a long involvement in Vietnam. Beginning with the OSS in 1945 and American equipment support to France, it was the fall of Dien Bien Phu, French disengagement and American underwriting of the new regime that placed America squarely in the middle of a conflict that was about to increase in tempo and intensity. For twelve years, from 1954 to 1965, American Armed Forces played an advisory role in Vietnam. In 1965 the political and military situation change dramatically, and the US intervenes directly into the conflict. During the period of 1965 to 1967 the US strategy was primarily strategic-defensive tactical offensive.[142] During this period, the military strategy was one that was characteristic of fire fighting. Units were positioned centrally to react to the border, rural and urban coastal threat. Additionally during this period, there were a number of operations, and a number of short, sharp engagements, such as the Ia Drang Valley, Landing Zone X-Ray, and Dak. For Hanoi, the same basic strategy existed. There was a direct relationship between the coastal regions and the rural areas. Hanoi sought to divert attention away from their cross-border incursions and urban area of support by initiating activities in the rural areas to draw away American forces. As Hanoi sought to relieve the pressure on units or forces that they felt were endangered, they shifted units and initiated action away from those danger areas. In essence, this was the same strategy that the Viet Minh employed against the French, and Communist forces employed against the South Vietnamese Army in 1964–65.

Tet should not have come as a surprise. The Communists took advantage of the 1966 and 1967 Tet truce to position forces and initiate offensive action as well as the 1966 Christmas Truce. By 1967 the Communists had numerical parity with American forces in country. The Americans could claim some modest victories and the Communists were getting worried over the technological and firepower superiority that American units enjoyed. By late 1967 the Americans felt that they were probably ready to go over to a major offensive.

The Communists pre-empt American plans for an offensive. The Tet offensive began in the middle of January 1968 when three

NVA divisions began to mass in the vicinity of Khe Sanh. While pressure and attention were focused on Khe Sanh, 85,000 Communist Troops prepared for Tet. [143] Other than Khe Sanh, some of the heaviest fighting during Tet occurred at Hue. Because of Hue's history and rich symbolism the Communists felt that Hue was the best place to gain a toehold that could be rapidly exploited. The Tet offensive was the embodiment of the use of the indirect approach and every aspect of "unrestricted" means set in a limited war setting. Hanoi did not have a timetable, just objectives. Additionally, American units were bound by a strict set of rules of engagement that the NVA never recognized. The NVA was never bound by any burdens or chains in the manner in which they conducted the war; on the contrary, they did a magnificent job exploiting the social and institutional asymmetrical differences.

Although it appeared to US intelligence that something was building, they had no specific indication that Hue was a target, that is, not until January 30 when a US Army radio intercept station at Phu Bai picked up traffic that indicated that Hue was a target. Regrettably, due to internal standard operating procedures the message was sent to headquarters in Da Nang for analysis. By the time someone decided that it was clear to forward on to Hue, it didn't matter anymore.

The actual combat for the city consisted of essentially, a two-phase fight. The first was for the urban and residential sprawl on the south side. Key and critical terrain on the south consisted of the MACV HQ's, the Treasury, the capitol, and the An Cuu Bridge. On the north side was the citadel. Within the citadel the key terrain was the Imperial Palace, the ARVN compound and the Tay Loc Airfield.

As the battle kicks off at about 0340 on the 31 January 1968, the enemy seemed to be everywhere at once. HQs through out the country were inundated with reports of activity everywhere. The NVA achieved complete strategic surprise. As the tempo of activity increases, no one could imagine the size, scope or magnitude of the offensive.

In Hue, the NVA had managed to assemble eight maneuver Battalions, two Sapper Battalions, and an undetermined amount of infiltrators.[144] This Division-sized force managed to control virtually the entire city within hours except for the ARVN and MACV HQs. For all intents and purposes it appears that the initial American and ARVN response is too little, too late. The NVA teach the Americans a lesson in asymmetrical doctrine as they manipulate both strategy and force. The NVA use unrestricted means and methods to exploit American values, ROE, and internal operating procedures that delay both the decision-making process and the dissemination of valuable intelligence. Command groups everywhere struggle and scramble to figure out just exactly what is happening to whom and where, essentially what is diversion and what is intended. As a result, nothing really tangible happens.

Ironically although the overall strategy in Vietnam was one of strategic defense, tactical offense, and although central positioning of units in the country was to allow American units to react to both the urban areas and border/rural areas, and that the previous two Tet holidays were followed by intense NVA offensives, there was no defense plan for the city. Additionally, American units were piecemealed into the fray over a period of several days and weeks. It appears as though the entire response was to rely on American flexibility, ingenuity and technological and weapons superiority. Supporting this premise is the X-ray staff at Phu Bai that directed LTC Gravel to assemble an ad hoc force from Alpha 1/1 and Golf 2/5, cross the Perfume River and link up with the ARVN HQs in the citadel. Certainly, this says much for the HQ's staff inability to fully recognize the gravity of both the situation and the decision that they just directed. But it also is indicative of the condescending attitude that Americans had for their enemy. In fine fashion of conflicts past, C3—command, control and communications—is directed from a remote location that had absolutely no real clue what was going on, and completely disregarded the advice that the local ground commanders offered. Within just a few hours Golf Company suffered fifty dead and wounded, or one-third of its total combat strength. Equally appalling was the refusal of ARVN tanks to cooperate with the Marines. Thankfully, LTC Gravel disobeys his orders when he sees the futility of the mission and breaks contact in order to return to the MACV compound. Adding insult to injury, as the Marines request help from the MACV HQ and try to evacuate the dead wounded, Col. Adkisson chose to ignore the request. Certainly there was a better way to handle that than complete disregard, even if Col. Adkisson was having some trouble of his own holding the compound.

On day two of the fight, overly optimistic reporting and misinformation continue to contribute to the debacle of the street fight. On day one, there were nine NVA Bn's in the city. On day two there were an additional five for a total of fourteen NVA Battalions. Overall, Tet is still in progress. Although the attack on the American Embassy is over, and some cities in the Mekong Delta have already fallen, fighting still rages in Saigon, Hue and Khe Sanh. As the NVA push the offensive, the Allies hamstring the strategic result by not allowing for the use of artillery, bombs or Napalm in Hue. The fight will be an infantry *tete a tete* brawl.

Amazingly, by day two, the NVA had almost complete control of the entire city with the exception of the two bastions of resistance—the MACV compound and the ARVN HQ. Additionally, Highway 1 was still open and the NVA had failed to either blow or completely secure the An Cuu Bridge.

By day three, units are still being piecemealed into Hue, and the fight turns to a house to house brawl where success is measured by one room at a time. It has been argued that the NVA took too long to seize the initiative. This is a valid point if we argue that the An Cuu Bridge over Highway 1 wasn't dropped by NVA sappers until day three. But then again, how could the NVA know the determination and resolve of the five Marine companies that managed to hold on in the city; two of which were able to cross the bridge and reinforce MACV HQ before the bridge was blown. The next day as the house to house brawl continues; American ingenuity, teamwork, firepower and camaraderie at the Company and Battalion levels compensate for poor decision making, piecemeal troop assignments, and political hamstrings that nullify combined arms operations. By the days end, success was another seventy-five yards gained.[145] Amazingly, through the entire battle, ROE and protocol would demand the attention of junior leaders and officers as they struggled for both survival and there sanity;[146] nevertheless, resistance on the south side of town essentially collapse's on the 6th of February when the Marines take the Provincial Capitol. The battle for the citadel would now begin and would take another, roughly, three weeks to complete. Essentially the fighting is daily from dusk to dawn with only a few brief pause's due to problems with re-supply.

By the 13th of February some of the restrictions on the use of artillery had eased and Army and Marine 8 Inch Howitzer batteries move in to support Hue. By the 20th all restrictions had been lifted and Phantoms are called in to strike targets in the citadel. Nevertheless, Americans were still plagued with problems in re-supply, ROE and protocol. Politics adds insult to injury when the Marines are told that the ARVN will storm the last bastions in the citadel and raise the Viet Cong flag. Ironically, the NVA had long departed and the ARVN attack through the citadel was largely a symbolic gesture meant to restore ARVN confidence and morale. By February 28 the official mop-up operations would begin.

All in all, twenty-six American units were cited for having participated in the battle for Hue. The Marine Corp reported 147 KIA and 857 WIA, for a total of 1004 men. This represents half of the effective combat power of all the Marines committed to the action in Hue. If this is true, then a total of about 2008 Marines were committed to combat or roughly the size of a reinforced regimental task force. This reinforced or regimental-size task force faced an enemy-sized division force of roughly six to eight thousand NVA.

In terms of analysis with current asymmetrical doctrine, the NVA achieved complete strategic surprise. Largely the success of this surprise was in the exploitation of known diplomatic and political weaknesses. Strategically, the Marines were restricted by protocol and ROE that delayed decision-making and reduced the overall effectiveness of combined arms operations by disallowing heavy support. Strategically, the NVA failed to exploit initial success by failing to eliminate the two pockets of resistance—MACV HQ and the ARVN compound. The Marines would exploit this and further strengthened their position by taking over the stadium and establishing security at the boat ramp and hasty LZ. Additionally they failed to secure and blow the Ann Cuu Bridge, and then later to the west, the second bridge was detonated by sappers after Marines had already secured a toehold. The US appeared to fail to gather and exploit twenty-two years of insurgent combat history. It is almost incomprehensible that the previous two years were also marked by major offensives at or after major holidays and essentially nothing was done. It appeared that the only lesson etched in stone was Dien Bien Phu and the result of this was a paranoia that still haunts the military today. Further, the NVA achieved operational and tactical success by training and preparing there combat units for MOUT. They specifically identified units that would be responsible for that city and focused their attention and resources on those units in order to insure success. Other strategic and operational factors that hindered operations were the relief in place of the ARVN airborne before the Marine's made actual physical contact and could sort through boundaries and tactical areas of responsibility on the ground. Additional factors were the failure of ARVN armor to support the immediate needs of Marine units engaged in close combat.

The actual brawl for the city was a classic infantry battle without the help or support of combined arms. Anti-tank, anti-sniper, anti-machine gun, and barrier buster weapons were improvised from 106 Recoilless rifles, Ontos, demolitions and copious amounts of hand grenades. Although the ARVN was essentially in sponsorship and under the tutelage of American advisors, it was expected that the ARVN would be able to defend themselves. They were organized under US style TO&E, and had US weapons and US training. The missing element was the social dimension of war. The US failed to calculate the effect of culture and ideology. The religious, ethical and ideological differences between Western and Asian cultures were so great that it was inevitable that the feelings between officers, men and units would span the spectrum of a lack of trust and understanding to outright derision and loathing. Although there were some exceptions to this rule, it was a rarity. Another factor was the unfortunate manipulation of the overall situation to the conventional strategy and tactics. In other words, the US attempted to form the situation to the strategy as opposed to forming the strategy to the situation.

The Marines bore the brunt of the fight. As usual, battalion and company-sized unit leadership, morale, teamwork and camaraderie would compensate for WWI-style decision making. Although it may be argued that the Americans failed to lose an engagement throughout the entire Vietnam War, we did come awfully close on a few occasions. Our redemption in this came a result of an overwhelming amount of firepower and technological superiority at the very last minute, when things became desperate and commanders finally captured someone's attention at HQ, and/or were willing to ignore the ROE for "danger close" and drop the arsenal on there own trenches, essentially, an attitude of, "If I am going to hell, then I taking a few of these guys with me." Nevertheless the NVA achieved their ultimate strategic and political aims. The NVA were prepared to sacrifice 85,000 men on a strategic objective. Both then and today, Americans would never tolerate the type of casualties that the NVA suffered to achieve strategic success. There are too many divergent groups that would exploit those incidents for there own narrow political objectives without considering the consequence of its future impact. The Clinton administration in Somalia is a classic example, but that is covered in the next chapter. In a quote from Col. Harry Summers, "it was an American tactical victory—strategic defeat and the lack of any serious military strategy contributions by the military in the post-WWII era."[147]

In comparison with current doctrine on asymmetrical warfare, asymmetry had been achieved through shaping and preparing strategies over fifteen years earlier. A combination of events conspired, or perhaps intended, impacted the mind set and decision making in sixty-five that resulted in the turning point in the war that began with Tet, Khe Sanh, and Hue. Some examples of the highlights that contributed to both US and NVA shaping strategy include: Eisenhower's refusal to intervene on behalf of the French; American Advisors in Vietnam; covert and overt diplomacy that advocated the overthrow of the Diem Administration; Kennedy's up-start response to the war by emphasizing the growth and greater use of special forces in a "whole new kind of strategy,"[148] an administration that recognized the need for "compound warfare" by advocating the training of regular Army/Marine units in counterinsurgency operations in conjunction with special forces; [149] and military leadership that was trapped in the specter of Korea and WW II strategies and were unable to modify the strategy to the situation, but rather attempted to mold the situation to a boiler plate conventional strategy. Lastly, the political agenda and response by Kennedy, Johnson and McNamara insured that the conflict would always be a low-key affair with out American popular support.

Although the Kennedy and subsequent Johnson administrations had done their homework and had read the writings of Mao, Fidel, and Che. They were also trapped in a situation that forced a direct and indirect confrontation with those that supported wars for national liberation. Flexible and limited response was not a valid option for the proverbial wars of "the people in arms." Although Kennedy, McNamara, and perhaps even Taylor had a firm understanding and vision of the new kind of war to be fought, military leadership was passive-aggressive and recalcitrant to this new way of thinking. Consequently, when the first units deployed to Da Nang in 1965 they were set up for failure.

In conclusion, there has been serious blame on the so-called intelligence failure, protocol, and strategy that effected and affected operations throughout the war. Arguably a ground commander should not have to wait two hours for the results of a request for fire mission. Units should never be piecemealed into an operation. There should be greater trust and cooperation between coalition partners. All of these macro-lessons from 1968 are as applicable today as they were during the Tet Offensive. As American Armed Forces head to the Middle East the disparity between Western and Eastern religions, culture, ideology and thinking are as wide as the Grand Canyon. It will take an extreme effort to bridge these social dimensions that contribute to the "fog of war." This "fog" will increase and impact operations exponentially unless steps are taken to work through and address these "training" issues jointly.

With Western war correspondents traveling with front line units, the "CNN Effect" will have as great an impact today as it did in 1968. The major difference is that technology has greatly compressed time and space since 1968, and collateral damage, accidents, incidents and atrocities will be both manipulated and widely misrepresented in an even shorter amount of time. Nevertheless, the strategic and operational failures were a combination of many elements that contributed to the debacle at Hue, beginning with the decisions made at the White House. The NVA were astute and enduring enough to concentrate on exploiting the political and social elements of the war. And lastly, they understood how to "shape and prepare" the battlefield "now" for the expected future success. Even if the fruits of the strategic and disproportionate effect would not be realized for four to six years after the events took place. They still achieved their strategic goals.

MOGADISHU

Mogadishu, Somalia, October 3, 1993, is the last tale of our trilogy of urban combat. This so-called modern day battle has many lessons for current doctrine and asymmetrical warfare. Once again, this tale of past history offers many sad lessons and comparisons in asymmetry that are very applicable today, and potentially, for future operations.

Sadly, the Somalia crisis has its roots in the late 80s. President Said Barre seized power in October 1969, and essentially established a socialist government with the aim of bringing about rapid social changes in a country that was locked desperately in the past. Armed opposition to Barre's government began in 1988 and he was forced to flee in 1991.

The Somalis are divided into six basic tribes or clans, and like their Islamic brothers in the Middle East, are associated with a defined territory. Clans are migratory and a person's loyalty and honor is to, and bestowed from, his sub-clan and tribe.

The internal fight for power in 1991 leads to massive conflict and famine. The two clans wrestling for power are Aideed's Habar Gidr sub clan and the clan of Ali Mahdi's Abgall. These clans fight for control of the ports, airports and UN booty as means of gaining absolute control. President Bush decided in 1992 that the US would secure the food distribution in order to reduce the growing famine. This part of the mission was successful. The failure begins internally when the Somalis fail to identify and accept a leadership team to represent all the tribes, hence trouble brews just below the surface. With trouble continuing to simmer just below the surface, Clinton declares success in May of 1993, but keeps US troops in the country as part of a UN mission that was required for "rebuilding Somali society and promoting democracy in a strife torn nation."[150] The mission of the UN force requires the installation and imposition of a government structure that the warring factions don't recognize. This UN-stated political objective clashes directly with Somali ideals and makes war highly likely.[151] Adding insult to injury, Clinton abdicates his responsibility as Commander in Chief to UN Secretary General Butros Gali. Although intervention may have been the right thing to do, the UN starts a war against Aideed that the Somalis see as a form of unfinished personal vendetta because of past bad blood between the Aideed clan and Butros Gali before Gali became Secretary General.[152] Additionally, the UN made no attempt to diminish the factional fighting and identify a single body of Somali representatives that could represent and govern the country. Lastly, they attempted a goal of nation building from a society that was still obscenely fragmented and cared more about increasing clan wealth, clan power, clan influence and day to day survival than long-term nation building.

As the UN decides that Aideed is the primary barrier to achieving their goals, the "war" essentially begins with the ambush and massacre of twenty-five Pakistani soldiers on June 5, 1993. As a result of that massacre, the UN declares Aideed a rogue and publicly declares for his apprehension and arrest. With in days, Ambassador Howe requests the aid of Special Forces to conduct counter-terrorist operations, declares a $25,000 dollar bounty and US troops begin attacking Aideed targets. [153] On July 12, as clan members gather in the Abdi house to discuss solutions to the UN issue, American Cobra helicopters attack the house, destroying it and killing a number of the clan gathered there. This one particular act is the straw that breaks the camel's back. For the Somalis, even those not associated with either clan, this is the deciding factor that the US and the UN are out of control and that something needs to be done. The Somalis seethe with anger and revenge for this callous use of force.[154] The remainder of the year is a continuation of a war in the shadows and the so-called low-intensity conflict begins exacting a higher toll as each month brings new attacks and reprisals.

Clinton makes the same mistake that LBJ made in 1965–68 and attempts to keep the whole thing low key and out of the press. The primary difference was that Clinton failed to build up force structure levels and force protection levels adequate to the situation and relied extensively on coalition/UN forces to provide essential support. In a comparison with the ARVN armor at the Battle of Hue, in September of 1993, Pakistani armor was used to support an American Engineer Company's road clearing operation in Mogadishu. As soon as they were attacked, the armor fled the scene causing the lightly armed engineers to fend for themselves and fight their own way out. The primary commonality in Somalia, Hue and Stalingrad was the questionable reliability of coalition forces and the lack of any real coalition mutual support and trust in each other, let alone the complicated task of coalition combined arms operations.

The Rangers arrive in August and they rapidly acquaint themselves with their tactical area of responsibility. Several missions result in success, however the Somalis become enraged at American audacity and ignorance as they view the use of helicopters as a means to intimidate and demean them. Additionally, the use of the helicopters in these daily low-level missions caused a number of cultural violations that bred and fomented anger at the Americans.

When Clinton allocated the Delta, Ranger, and 160th SOAR to the fight in Somalia he misallocated highly valuable resources. Arguably, the snatch and grab missions are a Delta/Seal and Ranger forte; however, a combination of many factors such as government policy, UN mission goals, lack of organic support, petty interservice power struggles, misunderstanding of the social elements of the conflict, and lastly, the underestimation, complacency and lack of respect for the enemy all contributed to the debacle that resulted in eighty-four WIA and eighteen KIA on that fateful day in October of 1993. Other factors such as remote controlled C3, a lack of training in urban combat, failure to adhere to equipment SOPs, unit personnel replacements not adequately trained and prepared for operations in country, dual command roles and responsibilities that violated unity of command, and a lack of overall trust and confidence in each others strengths and weaknesses—primarily between Delta and the Rangers.

In an attempt to keep the Aideed clan off guard, and the coalition forces that were feeding up to date intelligence to the Somalis, the strategy and missions had varied during the period of the Somali operation. Finally someone decided that in order to clean up this mess they should target "tier one personalities", a euphemism for targeting the Clan leadership of Mohamed Farrah Aidid. [155]

Tactically, the mission was a masterpiece of precision. The snatch and grab for the Somali clan leadership was organized around essentially, two groups: one air and one ground. The air element consisted of four each AH-6, four each MH-6 and eight MH-60, Black Hawks. In addition to this were three aerial observation helicopters and the Orion Spy Plane. The ground group consisted of nine Humvees and three five-ton trucks. The entire operation was a mix of Delta Squadron Operators, SEAL (sea, air, land) team, Rangers, the command group and the CSAR (combat, search and rescue) team, all in all, about one hundred and sixty men.

Strategically and operationally, the mission was a disaster of significant proportions. Politically and militarily, America had finally had its own Dien Bien Phu.

Arguably the mission was within minutes or seconds of being successful. If only that RPG had not taken out that first aircraft. On the other hand, it is easy to disregard the shaping, preparing and social dimensions of political and military strategy that had already shaped and prepared the battlefield against the US/ UN forces, particularly after the change of White House leadership from Bush to Clinton, and the subsequent policies that shaped the available military options. Unwittingly, the Americans contributed to a "War of National Uprisings."[156] Although the entire Somali operation was close to being successful, strategy, tactics, techniques and procedures helped turn the tide of general Somali support against them. Additionally, the normal amount of friction that is encountered in any operation was compounded by the cultural, leadership and organizational differences between all the tactical players/units that were required to work together.

There are a couple of sources for the definition of asymmetrical warfare; Field Manual 3–0, Operations has six short paragraphs devoted to the topic of asymmetry. These paragraphs talk about exploiting the disparity between organizations, structure, tactics, strategy and technology. NcNair Paper 62 has also proposed a definition, one that emphasizes exploiting the same basic elements; however it adds disproportionate effect with the emphasis on the psychological.[157] Arguably both the Somali's and Task Force Ranger attempted the same concept, each one focusing on those elements that would insure strategic success. Regrettably, the difference was in the US strategy misreading the social elements of this particular conflict and how those elements actually worked against us.

As we compare the asymmetry of operations and the guiding strategy of "shape, prepare, and respond now", the macro lessons for asymmetrical future operations become abundantly clear.[158]

Strategically: The social dimensions of planning the political and military objectives had been completely misread and/or disregarded. How could we just think that the elimination of the leadership would result in the collapse of a form of government—clan leadership—that is as old as antiquity? Examples throughout the Middle East and history abound with this failure. How could we not think that we would not create at least five more generations of rogues or terrorists with our actions?

Force protection structure and levels were not adequate to the environment. Even if we assume that Delta and the Rangers could have pulled it off, the lack of a combined arms force that relies on the interoperability and capabilities of the various weapon systems that provide mutual support and protection were not there. Not to mention an inadequate reaction force plan. Arguably, one might say that this is a conventional notion that does not apply to unconventional tactics. I would agree, however, the reliance on the UN commander and the 10th Mountain Division as a reaction force settles the argument in favor of this conventional notion.

The UN mission goals, coupled with the Somali perception of a personal and unfinished vendetta led by Butros-Gali under the protection and authority of the UN was a shaping factor that would never have been overcome.

The American forces had a terribly condescending attitude toward the Somalis, and completely underestimated there resolve. But rightly so, the situation was probably under control and might have been contained except for the exceptionally large blunder of attacking the Abdi house and killing clan leadership and innocent civilians that pushed the Somalis against anything and everything that the US and US forces represented.

General Garrison was a soldier's soldier, rather direct and to the point, with thirty years of experience in covert operations. He understood the strategy and the critical elements of strategy and he understood the operational and tactical execution. But how politically astute was he? Did he misunderstand the euphemisms and political obfuscation? Why did Colin Powell approve the plan without adequate force protection? What other officers would have resisted the administration in order to insure military success? Was career advancement and/or self-preservation the over riding factor? Was it possible that Garrison was being used as a fall guy? Or did these men really believe that a Battalion of Rangers and a few Delta Operators could cure the whole mess? What happened to the infamous in-depth psychological and country studies that Delta/SEAL and the Rangers performed? The entire premise of Special Operations is the intimate understanding of the psychological and social dimensions of the theater of operations and the effect on mission planning.

If the US and the UN felt that assassination or kidnap was a truly viable option, why did they allow the UN to announce its intentions and why did Howe offer a bounty for Aideed's head? Why didn't they offer to use diplomacy and politics in order to lull Aideed into a sense of complacency? This is almost as incredulous as the Russian effort at Stalingrad that attempted to authenticate the credibility of the German League of Officers. One laughs at the stupidity of its timing.

Operationally: Remote C3 gives a false sense of security and does not allow the command group to fully appreciate the situation. Arguably, someone needs to think at least two or three moves ahead, and someone needs to remain calm and level headed while the rest are about to lose there wits. But technology is also a chimera that has distinct disadvantages. Without the audio portion accompanying the video, there was no way a commander could get a real feel. Like the NODs, the remote C3 also reduces and eliminates the peripheral vision and the sixth sense or gut intuition that guides us during times of critically grave danger.

Differences in the way units operate and train, cultural bias, impatience, impetuousness, a sense of invincibility and the several recent events that psychologically shifted Somalis against the US were all variables that helped to create the conditions that were critical to an escalating friction that manifested itself during the tactical execution; six missions in six weeks and two incidents that caused political havoc and embarrassment created pressures that effected and affected the operation. The October 3rd raid, coupled with the previous missions, all caused collateral damage that increased spontaneous recruitment and bred a hunger for revenge. Lastly, implicit distrust between some of the commanders also filtered through and to the units. Once again this manifested itself during the course of the operation when Delta operators took formal charge or were seen by the Rangers as a more credible example of leadership. Arguably, when ones life is at stake, you are certainly going to follow the lead and example of the actual and perceived expert. One needs to speculate on how much larger or smaller the casualty count might have been if the Delta operators were absent from this action; certainly the Delta squadron bore the brunt of the casualty toll, but why? Besides the obvious examples of protecting the pilots, how many Delta casualties was a result of the operators attempting to guide the Rangers as well? Much cause for speculation here as well.

Tactically: Ranger leadership was certainly tested during this operation. Although the unit was highly cohesive and trained for combat, it was not necessarily for trained for urban combat. Additionally, the Ranger unit had not necessarily been battle tested and hardened; the Delta operators were and the troops knew that. Subsequently, as the action became more intense the Delta operators had a "referent" and "authoritative" power that none would have been able to transcend. Additionally, internal ROE and SOP should have been articulated a little clearer. Because the Rangers had supported the Delta in all its previous missions, and because the Rangers provided back up and security for the October 3 snatch and grab; it was patently clear that the Rangers were both in a support and subordinate role.

Arguably, the Rangers should have been practicing and training for combat in cities. Although the daring daylight raid was a snatch and grab, the Delta operators were the only ones that had any kind of SWAT/MOUT training. As a minimum the Rangers should have received some rudimentary training.

The Rangers and Delta, as would have been expected, did an excellent job at tactical improvisation of weapons, tactics, techniques and procedures. Arguably the learning curve was very high and they did it on the fly, regrettably at the expense of those that they witnessed dying.

This is one conflict in which the fog of war certainly lifted its ugly head. In some instances I would say that defeat was snatched out of the jaws of victory. With perfect hindsight we can speculate that a number of elements rearranged differently might have made the entire task force heroes instead of scapegoats.

Lastly, we might also speculate that it was the caliber of the unit and leadership that made the difference between a severe beating and a total catastrophic failure. How would the 10th Mountain Division or some other unit perform under the same circumstances?

In summary there are a number of lessons from the conflict in Somalia that are as applicable today as perhaps the future. Although Col. Kenneth Allard wrote *Somali Operations Lessons Learned* for the National Defense University under the auspice of the Joint Universal Lessons Learned System, it is primarily aimed at the JTOC and policy maker level, and is "politically correct" in its conveyance of potential challenges for future operations. The book does not offer much for a task force Ranger commander or those that would need more than the broad brushstroke offered by Allard. With that in mind there are nine lessons that can be offered to planners and operators for now and for future operations.

1. UN mission objectives are out of touch with ideological and cultural reality.

2. Misunderstood cultural and ideological values caused a miscalculation of the use of force on 12 July; as a result those undecided or neutral are no longer straddling the fence and shift support to the warlords. Additionally, this stiffens clan resolve.

3. As the escalation of low intensity conflict intensifies, did we underestimate the Somali clan strategy and military goals? It appears as though we did. Somali clan strategy was to increase casualties and public knowledge in return for a growing US and world opinion aversion to war, designed to force both withdrawl of forces and to force concessions at the diplomatic tables.

4. Clinton kept the whole affair away from both Congress and the US public. Any first year student in military studies knows that this is a cardinal sin that will ultimately backfire. It did!

5. Inadequate force structure, composition and organization. Absolutely no force protection was built into the organization.

6. Inadequate training in operations on urbanized terrain and an inadequate and ad hoc personnel replacement system that contributed to TF Rangers casualty rate.

7. Questionable reliability of coalition forces. Pakistani armor failed to properly support American Engineers. Equally disturbing is the fact that the Pakistani commander evaded his UN responsibilities when he kept his units out of specific areas of the city.

8. American policies, tactics, and strategy create the necessary ideological asymmetry that the US and the UN would never overcome

9. Policies shaped available military options and planning. We failed ourselves and our leaders failed us because we did not have the courage to question the application of force to this situation. Once again we attempted to mold situation to strategy instead of strategy to situation. Arguably, Howe's temperament was probably the leading causal to this debacle. Leader overconfidence and condescending attitudes certainly are no breeding ground for approaching enemy situations with a healthy respect of "humbleness."

In conclusion, besides the nine lessons listed above that could all apply to the cities studies in our trilogy, there are at least three asymmetrical macro trends from Stalingrad, Hue, and Mogadishu that are relevant to today's operations and concepts of asymmetrical warfare.

First, in all three battles the players misread the social dimensions of the overall strategy and in many ways contributed to their own thrashing by ignorance of the enemy resolve. This resolve came as an increasing condition as a result of the tactical and operational execution of the plan.

Second, in all three examples, the players had a very condescending and distrustful attitude of coalition partners. The enemy knew and understood this and exploited that weakness.

Third, the players did not take the time to train its units in the rudimentary aspects of combat in cities. Consequently, the learning curve was high, but at the expense of death and casualties. Regrettably both the Germans at Stalingrad and the Americans at Hue and the Mog were complacent about the men and training and sent elite, well-chosen units to accomplish tasks without so much as a minimum of training in MOUT.

Author photo of 3 of his associates checking out the low water crossing as they attempt to find a way to get to training at Camp Shir Zai after a rainstorm and flash flood

Author receives an award from an ANA Brigade Commander for excellence in training. This was one of many such awards while training the ANA.

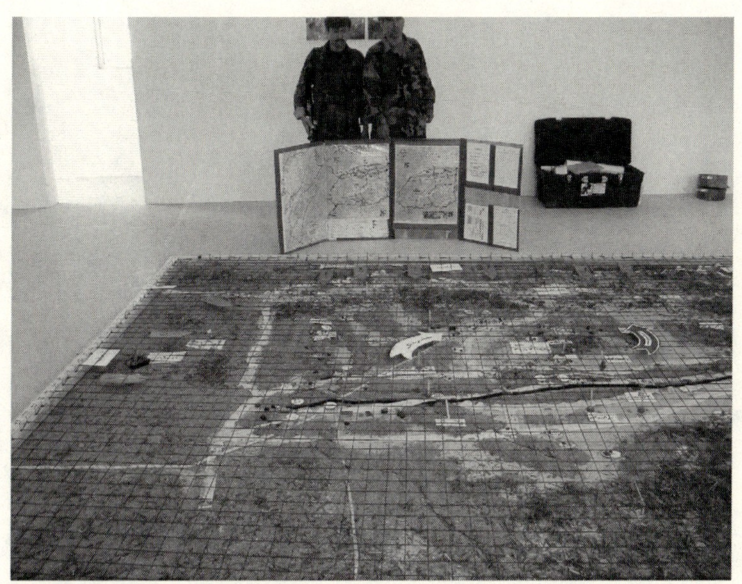

Two Afghan NCO's proudly stand next to the briefing chart and sand table that they prepared as part of NCO Battle Staff Training

A group of ANA NCO's and ANA police back brief the operation order along with the operational graphics using the sand table and briefing format taught in training.

Author photo of a group of children selling scarves and trinkets out in front of Camp Eggers. This is one small example of how children also helped support there families. These children also went to school for a couple of hours each day before selling there ware's near the Camp.

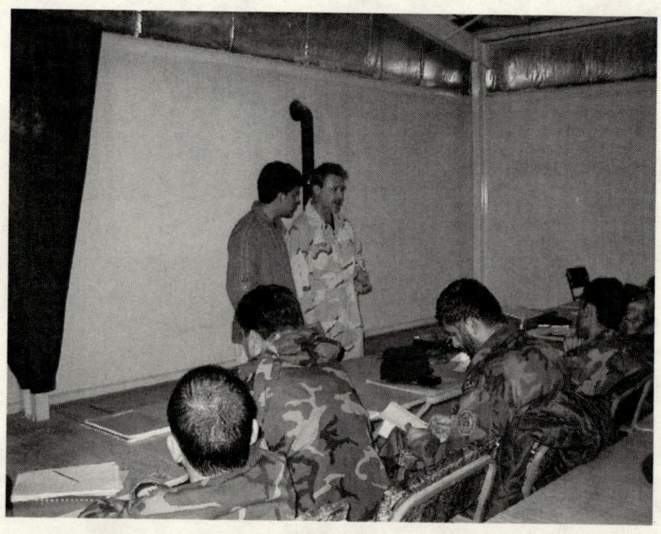

The author and his translator training one of many groups of ANA Officers and NCO's. Graduation Photo of Author and one of the Honor Graduate Students.

Photo of a small portion of Camp Shir Zai after a rainstorm

The Author and another American Trainer take a group photo of a group of ANA officers and NCO's in training.

The author with his ANA instructor counterpart and an American Advisor at the training center between classes.

Our translator, the author, class honor graduate and ANA instructor at a graduation ceremony.

The ANA Brigade commander passes out graduation certificates to the graduating class.

The Author and an ANA instructor participating in one of the many graduation ceremonies; passing out graduation certificates.

AFTERWORD ABOUT URBAN OPERATIONS

Urban operations are increasingly a part of conflict. Urban operations are also essential because of the many considerations that result in the decision to conduct an urban operation, for instance:

- Political considerations,
- Psychological considerations
- Operational considerations
- Potential center of gravity

The recent Israeli-Hezbollah conflict reflects at least two of the above considerations. Assuming that doctrine is the key element of training, that this training is the key to a units ability to execute an operation, and successful execution in turn results in a succession of operational successes, and/or one decisive strategic success, there are two predominate themes that are linked.

First the basic tactic's techniques and procedures (TTPs) remain unchanged: these are to suppress, breach, attack/assault and defend. Second, a defender and attacker make critical assumptions about the population when they make the decision to conduct urban operations, even if the urban area is a village or small town as opposed to a large city or metropolis. This decision is then executed by a commander and the size of the force to use is directly dependent on the commander's mission analysis. Regardless, the ultimate execution of the operation is done by a number of units, and the key factor is each squad, platoon and company and how well trained they are to conduct urban operations, which the training is derived from an understanding of doctrine, and an emphasis placed on that doctrine. Three key failures across the spectrum of urban operations are:

- complacency
- a lack of specialization of units and TTPs
- refugees and internally displaced persons

Cities are resilient and infrastructure continues to operate and remain intact despite conflict. The recent Hezbollah-Israeli conflict demonstrates that urban combat is still crucial and will be a center of gravity that will disrupt operations and will affect political and operational considerations. An analysis of urban operations between WWII and 1998 for urban operations in Stalingrad, Aachen, Manilla, Hue, Singapore, Beirut, Mogadishu, Grozny and Sarajevo, reveals a number of themes and lessons learned. These themes and lessons that are recurrent include: [159]

- complacency of your enemies resolve—Stalingrad, Mogidishu, Grozny (Iraq)

- overestimating your ability
- no contingency planning—Stalingrad, Grozny
- rubbling of the city—Stalingrad, Aachen, Grozny
- coalition soldiers operationally weak—Stalingrad, Hue, Mogadishu (Iraq and Afghanistan)
- high casualty rates
- misuse of armor—Stalingrad, Grozny,
- chicanery/deception/ruses (tactical)—seek to devise tactics and methods that put your opponent at a disadvantage
- ad hoc task organization—(Small mobile teams work best; 3 x 25 man teams, squad composition is also changed, no composite units
- improvisation of weapons and tactics (hugging, weapon for multiple uses, i.e. RPG's or Recoiless Rifles)
- infantry team re-organizations
- specialized urban training and specialized training to work with armor.
- published doctrine is of little help except for pointers on defense
- methodical and firepower intensive operations
- battlefield is not linear and clean

- snipers, mines and communications are your most important assets
- urban structure and architecture is important
- ROE/ROI
- no security
- no logistics plan
- information/PSYOP war: discrepancies between official sources and news media (public opinion is extremely important; prepare provocations to destabilize along ethnic lines—Grozny, Israeli-Hezbollah, Iraq, Afghanistan)
- extensive use of ECM and satellite blocking
- assaults uncoordinated; lack of coordination between units, agencies and police.
- technology is ineffective when the soldier or user is untrained, confused and afraid to use it.
- the capacity of clans, culture and social traditions goes beyond surface discipline
- force ratio's not adhered to
- *the decision to fight in successive cities*
- preparation for urban combat has to begin in peacetime

Increasingly, urban operations will also be a requirement of counterinsurgency. Generally speaking the area known as the

Middle East is comprised of eighteen nations, is 4.3 million square miles, had a population of 326 million in 2002, and 57% of this population resides in urban areas. For comparison, 25% of the population was urban in 1960 and current projections indicate that by the year 2015 that 70% of the population will be urban.[160] Increasingly, military operations and/or counter-insurgency operations within urban areas means that operations will increasingly include a succession of urban operations that literally will cross ethnic and cultural boundaries from one street to the next and will increase the complexity of operations.

Of particular importance is the fact that Middle Eastern cities are unique and different from Western and European urban design. Despite the fact that some Middle Eastern cities are changing and may begin to resemble Western design as a result of sprawl and growth, the city center in Middle Eastern cities still has a significant religious, political and social influence, and Middle Eastern urban design is also a physical reflection of tribal, clan and community separation that only the locals will know and understand, thus adding another layer of complexity to the hearts and minds campaign.

OVERALL CONCLUSIONS AND SUMMARY

Although the concept of airmobility and the use of helicopters began in the 50s, it was fully developed during the Vietnam War, and the infamous LZ X-ray represents the culmination of that effort. Then, like now, military change during a time of conflict sought to enhance our warfighting edge by combining technology, doctrine, organizational change, training and strategies to gain an asymmetrical edge.

Transformation or change also took place in the 70s and 80s with the development of Air-Land Battle doctrine, a doctrine that sought to synchronize in a "deep battle" the assets required to interdict, disrupt and destroy support units, command and control and follow on echelons, so the combat units fixed in the forward edge of the main battle area could focus on destroying those lead elements without worrying about the reinforcing and exploitation elements. A commander sought an asymmetrical advantage through synchronization of the battle space; in essence the precise and artful timing of the science of war and the enduring capabilities that each type of unit brings to the battle space. Air-Land Battle concepts were vindicated before a world audience in 1991 when the US Army and coalition forces literally annihilated the Iraq Army in a one- sided battle that veterans of this conflict derisively call the "100-hour Live Fire Exercise."

In the mid-90s Air-Land Battle concepts remained in place, however, the term Air-Land Battle was now replaced with "rapid dominance," and the US Army sought to cover its bases by emphasizing operations other than war, essentially a catchall for everything that the US Army might find itself involved in from humanitarian operations around the globe to security and stability operations in Bosnia.

During the period 1991 through 2001 the Army understood that the world situation was changing but the perception of an "imminent threat" disappeared with the collapse of the Soviet Union. The military recognized that it was the dominant military power and that it was probably going to have to be involved in a number of operations other than war; however, there was no real impetus for change because there was no longer a perceived substantial threat to American security. The global war on terror ushered in a new threat; asymmetrical threats in the form of an insidious ideology and from radical non-state actors sponsored by the resources and financial backing of states. It has taken the US military five years to begin to adapt to this so-called new form of war, yet it has done so in convoluted fits and starts. Perhaps what are new are the size of the area of operations and the complex number of additional players. Insurgency usually encompasses a couple of states; our asymmetrical war on terrorism is "global" and these global difficulties can be seen through the lens of the conflict in Iraq and Afghanistan, to a lesser visible degree in Malaysia, and to a very large degree through the UN.

Doctrine, strategic aims, training and ability are intrinsically connected. Failures in analysis and perception are major causes of miscalculation that result in an accumulation of failure and/or immediate catastrophic failure. Consider for example the following statements:

- rejection of tutelage and influences of the governing regime

- anarchism appeals more than self help

- governance at an emotional distance with a clearly immutable difference that despises anything less than competence measured in western terms.

- ethnic groups with racial tensions live side by side in communities yet separated with very limited to no interaction

- staunch tribalists, but believe that they will be worse off under a different regime, hence, the shifting attitudes and coalitions based on the perception of who has the power to administer swift justice or hefty amounts of carrots

- long-established patterns of rivalry that bubble intensely just barely below the surface

- pseudo-arguments of unfitness for combat

- varied relationships that at once encourage a partnership yet also encourage looting and theft

- You must be one or the other: with us or against us.

If these statements sound familiar to Iraq, Afghanistan and the Global War on Terror, is it just a coincidence? These statements describe the overall relationship of the British military and colonial government with there subjects in the Imperial British Crescent of Southwest Asia in 1941.[161]

The spirit and essence of counterinsurgency must primarily address a number of macro variables that include:

- the actual and active fighters; this is approximately 20% of the total problem that ultimately effects the remaining 80%

- active supporters

- the people; the population is both key and critical terrain, and includes the population within your area of interest and your area of operations

- sympathizers; sympathizers must be influenced to shift allegiance

- global opinion and information operations; you must be proactive and timely in managing global opinion.

Counterinsurgency is approximately 80% hearts and minds, and 20% combat operations. In essence counterinsurgent or insurgent doctrine must work to influence and own the population, the enemy's forces and global opinion. Commanders, strategists and politicians must work to insure a unity of effort and mindset by emphasizing small unit operations at the company team/task force level and below, by allowing for wide latitude of execution within centralized, synchronized and coordinated plans. This synchronization is with other military units, government agencies and non-government agencies to insure that the appropriate balance of hearts, minds, humanitarian, information, and combat operations is effective and all working towards the same goals established by the national authorities.

In essence, current doctrinal principles for conventional war and insurgencies are fundamentally sound. Integrating conventional doctrinal methods with counterinsurgency or asymmetrical warfare is not always necessarily wrong, but equally is not always necessarily right, and flexing or adjusting techniques, tactics, procedures, and methods should be considered within the context of the contemporary operational environment. Primarily, this means that decisions should be weighted heavily towards capabilities, that is, individual capabilities over rank. It is glaringly obvious that the intangibles and decisions derived from the consideration, or lack of consideration, of these intangibles cannot be overcome with either decisive force nor can one effectively achieve its strategic aims.

A recent example of this mismatch can be found in the recent Israeli-Hezbollah conflict. For instance, the two-week air campaign, the belated and limited use of ground force, a miscalculation of resolve and political potency, and the increasing effect that the global media has on decision makers in both the military and government. This is also a poignant example of the fact that your local on-the-ground public affairs officer and information officer need the ability to fight the information war by making on the spot decisions and announcements without clearing it first with higher headquarters.

Another key factor is unit specialization. The Army prides itself on the ability to adapt to all possible contingences and to react to a multiple set of operations that fall within the category of operations other than war, but this is a fallacy. Insurgency and counterinsurgency are special operations that require special training and a special mindset. This is not so much a large doctrinal shift as it is a very large mental and mindset shift. As an example, during the Falklands war, some British units, like the British paras had units deleted from deployment because of the investment in counterinsurgency training. In essence these units were trained for deployment to Northern Ireland and this training was considered an investment that could not be wasted on a conventional operation.

Traditional force multipliers such as leadership, morale, resilience, and belief in the justness of the war are important intangible variables that can change the entire outcome of battles and war. Commanders sometimes make critical tactical decisions regarding employment, effects and capabilities based outcomes on these intangibles when they specifically select a Ranger battalion, special operations unit or tier 1 conventional unit like the 82nd Airborne or 101st Air Assault to lead the so-called "Tip of the Spear." Commanders are unequivocally counting on intangible force multipliers to insure the outcome of an operation and the validation of doctrinal concepts; counterinsurgency operations that utilize these type units and/or so-called "conventional units" however requires a special emphasis on training and skills that to some small degree touch on the peripheral of military studies, but are not considered strategic enough to warrant special training, such as in: economics, diplomacy, conflict resolution and humanitarian operations.

Intelligence indicators and psychological warfare, also intangibles that are important in the decision-making process, should also be orchestrated in consonance with the select application of select approaches and should take priority over current conventional notions of capabilities or effects based operations, which in effect discount less quantifiable and analytical methods that appear to be less than "decisive" in the immediate expectation of results.

In essence a commander flexes a response to an operation based on his initial estimate of the situation within the parameters set by the political leadership and within the context of the cultural, political and social expectations and conditions. Further integrating and applying doctrinal methods and response is also heavily dependent on the culture and attributes of its army and its nation. Sometimes the change required to adapt to lessons learned is not always easily recognized and may require a paradigm shift of monumental proportions.

EPILOGUE

Doctrine provides a common language and framework in which organizations conduct operations. To examine doctrine also requires an examination of a myriad of other variables, variables that determine the successful application of doctrine and more importantly, success in conflict. Learning organizations "learn" to fail fast, evolve and change and create new doctrine, tactics, techniques and procedures to adapt to the environment. It is a "living" process that is constantly evolving as players and actors in the conflict learn to innovate on the fly and out fight each other.

Current doctrine may not require a radical change to adapt to the 21st Century; however, if we fully expect warfare to continue to move from conventional to insurgency/counterinsurgency then a radical shift in thinking, training, and organizing for combat will require immense change. It will also require a fresh look at how we use and apply the principles and tenets of war. In the last several years that I have spent in the Middle East in Saudi Arabia and Afghanistan, I have seen no such shift in thinking in the way the principles, tenets and doctrine might be applied. I have seen no such shift in thinking in either the military leadership or the combat units that compose the bulwark of our commitment to winning the Global War on Terror. The caveat to this statement is that they have adapted well to the tactical techniques and procedures, however, have not made the required mental and cultural shift. We have created drills and shared knowledge on countering IEDs, improving cordon and search, sensitive site exploitation, sniper/countersniper techniques and a host of other survival drills that improve survivability. We still have not made the most important shift of all—the mental shift from conventional operations to counterinsurgency. Security and stability operations is still largely viewed as a necessary evil, not my job, and should never include any overlap in duties, functions or responsibilities that might remotely be considered "non-military" in nature. In other words, to use a current military term, anything non-kinetic in its response is the last considered option.

The Army has expressed a desire for its soldiers to become *pentathletes*, soldiers that can adapt well to a myriad of environments. This implies an expertise in doctrine, culture, training and operations of a very finite nature; and a level of perseverance that is second to none. By definition it means that this athlete/soldier must make radical mental shifts as they move through each different event of the pentathlon. It takes years for an athlete to prepare for a pentathlon, which explicitly means a very specialized training regimen. By implication this is contrary to current US doctrine, which still advocates a generalist approach, and to be prepared to fight in all possible contingencies; contingencies or operations other than war that do not mean war fighting.

The Army has officially released its new counterinsurgency manual, but despite the accolades, one gets the distinct impression that, like the original 1976 version of FM 100–5, it does not necessarily inspire confidence in the force and will also require the US Army to think, train, act and organize differently. How does one reconcile the perceived lack of progress in fighting the war on terror? Can this lack of progress be attributed to doctrine, training, organization, institution or all of the above? Or is this perception of American media self-fulfilling prophecy? What tools now enhance war-fighting? What are the non-kinetic skills that are required at the small unit level?

In conclusion, there is a plethora of doctrine that the US Army has that routinely collects dust. It is a standard US military proverb that is usually quoted with pride to quote an unnamed Russian source that said that the most perplexing thing about the US military was that it did not follow its own doctrine. Doctrine, with some modifications is fundamentally sound for the 21st Century; what requires revision is the principles, tenets, mind-set, culture and organization if the US Army truly wants to cultivate an army of pentathaletes. The US Army has finally released its new counterinsurgency manual, it will remain to be seen if they can make the mental paradigm shift required to implement this new doctrine or will it remain mired in its old ways.

ENDNOTES

[1] US Army FM 3–24, Counterinsurgency, December 2006, US Department of Defense, pp IX-X

[2] US Army, FM 3–0, Operations, dated June 2001, preface

[3] US Army FM 7–0, Training the Force, Oct 2002, page iv

[4] Authors observations of CSTC-A TAG IV and V. Despite an attempt to write evolving doctrine, the US Army's bunker mentality negates its understanding and ability to execute tactical operations with the strategic aim in mind.

[5] See call handbook NO. 04–16, Cordon and Search, dated Jul 04; other references include CALL Handbook 02–8, June 02; CALL Handbook 03–35, Dec 03; and CALL Handbook 05–06, Jan 05. See also FM 1, *The Army*, June 2005, pp 1–20,1–21; lastly, the author was personally responsible for writing and adapting these US Army doctrine tactics techniques and procedures to the Afghan National Army.

[6] The primary doctrinal reference for an Infantry Company or unit performing Infantry like operations begins with FM 7–8. The corresponding Drill

and Training and Evaluation Outline manual is FM 7–8 Drill and FM 7–8 ARTEP.

[7] While the author was writing doctrine and acting as a trainer/mentor to the Afghan national Army; French and British Units were assigned to KMTC (Kabul Military Training Center) to train the ANA in US Army doctrine in the ANA Officer Candidate Course, Basic Training and NCO Course. Although we adapted an Officers Handbook that was based on some British Techniques, there was no exchange of doctrinal lessons learned within the institution outside of what the US Army or MPRI deemed appropriate. In other words all doctrine was primarily TTP oriented. Additionally, RTAG IV, Col Eggers,, asked MPRI to produce visual doctrinal aids that contradicted current US Army doctrine because according to US Army trainers, the ANA was unable to perform the TTP of Traveling Overwatch according to current US Army Drill standards as described in FM 7–8 Drill.

[8] See *CRS Report For Congress, US Army's Modular Redesign: Issues for Congress, 10 Feb 2005;* this also includes my personal involvement as a military contractor in both Afghanistan with CSTC-A (Combined Security Transition Command-Afghanistan) and with OPM-SANG (Office of Program Management-Saudi Arabian National Guard) lastly, see seminar transcripts of the 2005 Military History Symposium, An Army at War: Change in the Midst of Conflict; sponsored/hosted by TRADOC and the US Army Combat Arms Institute, Ft Leavenworth, KS.

[9] FM 1, *The Army*, June 2005, pp 1–10 and 1–12

[10] Airmobility in Vietnam, 1961–1971, pgs 3–8

[11] Ibid page 8–9

[12] Ibid

[13] Ibid page 20; Secretary McNamara, much like Donald Rumsfeld of today, had serious reservations about the Army's ability to produce a reexamination of the transformation or modernization concepts that would produce fresh, unorthodox concepts. See page 19 of *Airmobility, 1961–1971*

[14] Ibid

[15] Ibid page 22

[16] Authors Italics

[17] Ibid, page 20–24

[18] The source documents for this are the US Army, Center for Military History; Seven Firefights in Vietnam and Airmobile Operations. Increasingly, the US Army used helicopters in its missions and mission support. The largest air-mobile operation in the early years occurred in June of 1964 with the airlift of 1300 Vietnamese Marines. In June 1965, 2000 Vietnamese marines used helicopters to Air Assault positions. The 1/7th Cavalry, 1st CD would conduct search and destroy mission in the Ia Drang less than 6 months later in Nov 1965. In the words of CMH;1/7 Cav was going to Air Assault in to "develop there targets" based on intelligence estimates. The new Air Assault techniques gave them a "quick strike" capability. Airmobility operation was adapting the use of conventional infantry tactics and was primarily a weapons platform and a movement platform.

[19] Deciding What Has To Done: General William DePuy and The 1975 Edition of FM 100–5, Operations, Maj Paul E. Herbert, Leavenworth Paper Number 16, US Army Combat Studies Institute, US Army, Command and General Staff College, Introduction, pp1–2

[20] Nixon or Guam Doctrine reduced American Strategic planning from a 2 ½ war concept to a 1 ½ war concept. In essence the doctrine of the 1960'S planned on 2 simultaneous wars in different regions and 1 minor war. The Nixon Doctrine changed this to 1 ½

[21] Ibid, pp12–13

[22] From the introduction to *The Roots of Strategy*, Book 4

[23] FM 1, *The Army pp 1–20 and 1–21* and FM 3–0, *Operations,* p 1–14

[24] FM 1, *The Army*, June 2005

[25] My personal account of this comes from several meetings with US Army officers of TAG IV that are charged with ANA Institutional Training. Training that is supposed to build the foundations of a standing army yet prepare them

for counterinsurgency and SASO Operations. It is clear that teaching "Security and Stability Operations", which is nothing more than a concept that is contained in no less than 7 different manuals (FM 3- 06.11, FM 3–07, FM 7–98, FM 3–24, FM 90–8 and FM 3–07.22, FM 100–20) is still treated as a stepchild core curriculum that will eventually be replaced with more conventional notions. *COIN and SASO are still treated as conflict and doctrinal anomalies.*

[26] US Army, FM 3–0, *Operations*, dated June 2001, page 1–2

[27] John Nagl was featured on a CSPAN interview on 10 September 2007 in an interview with Sean Naylor of the Army Times. The topic of the interview was the new counterinsurgency manual.

[28] FM 3–07.22 *Counterinsurgency Operations*, dated Oct 2004, page vi

[29] Ibid, page 1–13

[30] FM 2–0, *Intelligence*, page 3–1 addresses offensive decisive operations as part of Full Spectrum Conflict. See also FM 1, *The Army*, dated, June 2005, page iv, The Soldiers Creed, and page 1–1 and page 1–2. Besides my personal experience of 24 years in the military; these are but two small examples that serve to illustrate the idea that *Decisive Operations* are culturally tied to military culture despite the type of operation that the military may be called to engage in.

[31] *Revolutionary War*, John Shy and Thomas W. Collier; *Makers of Modern Strategy , From Machiavelli to the Nuclear Age*, ed, Peter Paret, 1986

[32] Ibid

[33] See US Army FM 1, The Army

[34] Regrettably with this many manuals there is some confusion, and more often than not, based on my personal involvement in doctrine, the high level debates continue to discuss whether we have morphed from a counterinsurgency to SASO despite the still current high level of violence in Afghanistan.

[35] Conversation with Skip Both, LTC, 5th Group. In essence they know what skill sets they need and they can identify them when they see them, how do you teach that at an institutional level across an Army?

[36] See Dr. Colin Gray, *Irregular Enemies and the Essence of Strategy, Can America Adapt?*, US Army Strategic Studies Institute, March 2006; and *Changing the Army for Counterinsurgency Operations*, Brigadier Nigel Alywin-Foster, British Army, Military Review, Nov-Dec 2005. A response to Brigadier Alywin-Foster's article appeared in the March-April 2006 Military Review, *OIF Phase IV: A Planners Reply to Brigadier Alywin-Foster; Regrettably this reply was nothing more than an attempt to vindicate the US Army and did not adequately address the larger issues of essential transformation.*

[37] See Dr. Colin Gray, *Irregular Enemies and the Essence of Strategy, Can America Adapt?*, US Army Strategic Studies Institute, March 2006; and Changing the Army for Counterinsurgency Operations, Brigadier Nigel Alywin-Foster, British Army, Military Review, Nov-Dec 2005. A response to Brigadier Alywin-Foster's article appeared in the March-April 2006 Military Review, *OIF Phase IV: A Planners Reply to Brigadier Alywin-Foste; Regrettably this reply was nothing more than an attempt at vindication and the author failed to grasp the larger critical elements that needed to be addressed.*

[38] Personal experience with the ANA 1st Bde 205 Corps and 1st Bde 209th Corps. While I was training Officers and NCO's to prepare for combat operations the assigned ETT was ill-prepared and unknowledgeable of the essential tasks that needed to be trained. In several instances I was training the ETT while I was training the ANA.

[39] This was a specific mission task that I received from senior leadership when they confided that the leadership was not confident in the ETT's knowledge or ability to execute the training mission. This was further reinforced when ETT's attended my ANA field training classes and I had to Instruct them in basic fundamentals doctrinal products for Contract Year 2006–2007 Additional unstated work also included revising FM 7–8, Chapter 4 and 6 and numerous video illustrations of Sqd and Plt Drill/TTP's

Intelligence Manual	8	Tucker	Ishan	Kandak and below
React to IEDs	2	Cody	Hekmat	SQD & PLT
Hasty Traffic Control Points	4	Morris	Ishan	SQD & PLT
React to Contact & to Ambush, Mounted	1	Cody	Gogar	SQD &PLT
Sensitive Site Exploitation	6	Tucker	Shen Gul	Company
Critical Site Security/FOB Security	7	Morris	Nadar	Company
Cordon and Search	3	Tucker	Shen Gul	Platoon and Company
Individual and Small Leader Handbook	9	Cody	Hekmat	Individual
CQM &CQB	5	Cody	Nadar	Individual
Clear a Cave	10	Cody	Gogar	PLT & Company

EVOLVING DOCTRINE
ANA?, Intelligence Manual
Cordon and Search at PLT Level (Offensive Operation addition to existing ANA 7–8)
Sensitive Site Exploitation (SSE) at PLT and CO levels (Offensive Operation additions to existing ANAs 7–8 & 7–10)
React to IEDs at SQD and PLT levels (Battle Drill addition to existing ANA 7–8-Drill)
Tactical Convoy Operations, specifically, **React to Contact, Mounted** at SQD and PLT levels, and **React to Ambush, Mounted** at SQD and PLT levels (Sustainment Operation additions to existing ANAs 7–8 & 7–10). Includes load plans for Ranger Pick-Ups (see Small Unit Leader Handbook below)
Clear a Cave at SQD thru CO levels (Offensive Operation additions to existing ANAs 7–8 & 7–10)
Hasty Traffic Control Points at SQD and PLT levels (Battle Drill addition to existing ANA 7–8 Drill
Critical Site Security/FOB Security at CO level (Sustainment addition to existing ANA 7–10)
INDIVIDUAL AND SMALL UNIT LEADERS' DOCTRINE
Small Unit Leader Handbook. This is a stand-alone product, pocket sized, soldier-proofed, that contains checklists and blank formats for field orders, PCCs, PCIs, vehicle load plans, commo equipment and procedures, call for fire

Draganov Sniper Rifle Fundamentals and Marksmanship. Another of a continuing series of marksmanship pamphlets that will be published as a stand alone pamphlet plus will be integrated into the CTT Pam.
Basic Demolitions. Another of a continuing series of marksmanship pamphlets that will be published as a stand alone pamphlet plus will be integrated into the CTT Pam.
Close Quarter Marksmanship (CQM) and Close Quarters Battle (CQB). Another of a continuing series of marksmanship pamphlets that will be published as a stand alone pamphlet plus will be integrated into the CTT Pam.

[40] With the very recent exception of FM 3–24 that became official doctrine in Dec 2006

[41] The author spent 1 year in Saudi Arabia and one year in Afghanistan writing and training COIN Doctrine. There is a huge disconnect between those that are doing the planning and those that are doing the training. Worse is the fact that the ETT's which consist of Guard and Reserve personnel are ill equipped and lack the knowledge to integrate a tailored doctrine for the ANA and training.

[42] *In Praise of Attrition,* Ralph Peters, PARAMETERS, 2004, PP24–32; *Irregular Enemies and the Essence Of Strategy: Can the American Way of War Adapt?,* Colin S. Gray, March 2006, Strategic Studies Institute, US Army War College, Carlisle, PA; *Toward Combined Arms Warfare: A survey of 20th Century Tactics, Doctrine and Organization,* CSI Research Survey Number 2, Jonathan M. House, US Army Command and General Staff College, FT Leavenworth, KS, Aug 1984: *CRS Report For Congress, US Army's Modular Redesign: Issues for Congress, 10 Feb 2005*

[43] Ibid

[44] Ibid

[45] Ibid

[46] Ibid

[47] A General's New Plan to Battle Radical Islam, Greg Jaffe, Wall Street Journal, 2 Sept 2006.

[48] Analysis: US Unsuited for a long war, Pamela Hess, UPI, Aug 31, 2006

[49] I have relied heavily on the monograph, Recognizing and Understanding Revolutionary Change in Warfare: The Sovereignty of Context, Colin S. Gray, February 2006, Strategic Studies Institute, US Army.

[50] Maurice Matloff, *Allied Strategy in Europe, 1939–1945*, page 677, Makers of Modern Strategy from Machiavelli to the Nuclear Age, ed, Peter Paret

[51] Ibid, *Strategic Planning for Coalition Warfare*, 1940–1941. *Strategic Planning for Coalition Warfare*, 1942–1944; *Command Decisions*;

[52] Gole, Henry G. *The Road to Rainbow*: *Army Planning for Global War, 1934–1940, Naval Institute Press, 2002*

[53] Maurice Matloff, *Allied Strategy in Europe, 1939–1945*, page 678, Makers of Modern Strategy from Machiavelli to the Nuclear Age, ed, Peter Paret

[54] Gole, Henry G. *The Road to Rainbow*: *Army Planning for Global War, 1934–1940, Naval Institute Press, 2002*

[55] Maurice Matloff, *Allied Strategy in Europe, 1939–1945*, page 679, Makers of Modern Strategy from Machiavelli to the Nuclear Age, ed, Peter Paret

[56] Ibid

[57] Similar arguments might be made in comparison with the GAZA pullout

[58] And the Bush Administrations seemingly cowboy irreverence and indifference to European Allies

[59] Gole, Henry G. *The Road to Rainbow*: *Army Planning for Global War, 1934–1940, Naval Institute Press, 2002*

[60] 9/11 Commission?

[61] 1972 National security advisor memo on a FP for the Middle East

[62] Weigly, Russell, *The American Way of War: A History of United States Military Strategy and Policy*, Indiana University Press, 1977 Echevarria, Antulio

J, *Toward an American Way of War,* Strategic Studies Institute, US Army War College, March 2004

[63] *On War,* Carl von Clausewitz, ed, Michael Howard and Peter Paret

[64] *On War,* Carl von Clausewitz, ed, Michael Howard and Peter Paret, page 55,56,72

[65] The moral aspects referred to here are from On War as edited and translated by Michael Howard and Peter Paret and a commentary by Bernard Brodie

[66] McNair Paper 62, Chapter One, page 2

[67] Author speculation and personal interpretation from media op-ed and news release from FOX, CNN, MSNBC, Kuwait Times and Arab Times

[68] Maj Robert A Doughty, Leavenworth Paper Number 1, *The Evolution of US Army Tactical Doctrine 1946–1976,* CSI/CGSC Press; Jonathan M House, Research Survey Number 2, *Towards Combined Arms Warfare: A survey of 20th Century Tactics, Doctrine and Organization,* CSI/CGCS Press;

http://www-cgsc.army.mil/carl/resources/csi/csi.asp#cgsc

FM 5–0, Army Planning and Orders Production, 2005; FM 100–5, Operations, 1976, US Army; TRADOC PAM 25–34, Desk Guide to Doctrine Writing, 1992, US Army; FM 3–0, Operations, 2001, US Army

[69] Record, Jeffrey, *Bounding the Global war on Terror,* Dec 2003, Strategic Studies Institute

[70] US Army, FM 3–0, Operations, Chapter 4–1, 4–3 and Chapter 6; Antulio J. Echevarria II, *Clausewitz Center of Gravity: Changing Warfighting Doctrine Again*, Sept 2002, US Army War College, Strategic Studies Institute.

[71] Ibid

[72] Antulio J. Echevarria II, *Clausewitz Center of Gravity: Changing Warfighting Doctrine Again*, Sept 2002, US Army War College, Strategic Studies Institute.

[73] Ibid

[74] This is a reference to the original intent of the Air-Land battle doctrine and its subsequent evolution and validation during the Gulf War of 1991

[75] Thibault, George, E; Pg 2, pg30–35, *The Art and Practice of Military Strategy*, National Defense University, 1984

[76] Thibault, George, E; Pg 2, pg30–35, *The Art and Practice of Military Strategy*, National Defense University, 1984

[77] Boyd, John R; *Patterns of Conflict* pg 138–143; http://www.d-n-i.net/fcs/boyd_grand-_Strategy.htm

———*Strategic Game* pg 53–57

[78] Ibid

[79] *A Case Study in Fourth Generation Warfare: The "Al Aqsa" Intifada*, Dec 2000: http://www.d-n-i.net/al_aqsa_intifada/index.htm

[80] This triad consists of FM 1, The Army, FM 3–0, Operations and FM 7–0, Battle Focused Training. Arguably, some would say it also includes FM 5–0 Army Planning and Orders Process.

[81] Letter, Curtain to MacArthur, 7 Apr 1943, with 10 Apr reply. 1061/43; memo BG B.M.Fitch to Allied Commanders, "Air Attack of Objectives, within the Philippine Archipelago", 1 Nov 1944, and message from same to MacArthur , 2 Sept 1944, 11061/41, Records of US Army Commands, RG 338, National Archives

[82] Sean Naylor, *Not a Good Day to Die, The Untold Story of Operation Anaconda,* Penguin, 2004: see also General Tommy Franks, *American Soldier*, Harper Collins 2004, pp 377–381

[83] *Whitehouse, Pentagon to Release Interrogation* Memo's, CNN, 22 June 2004, http://www.cnn.com/2004/ALLPOLITICS/06/22/rumsfeld.memo/index.html

Rather, Dan; Rumsfeld *War Plan Criticized* , 4/1/2003 http://www.highbeam.com/library/doc0.asp?docid=1P1:73026772&refid=ink_d5&skeyword=&teaser=

Nation at War: Under Fire; Rumsfled's Design for War Criticized on the Battlefield, New York Times, Bernard Weinraub with Thom Shanker 4/1/2003

http://query.nytimes.com/gst/abstract.html?res=F60714FB3B5D0C728CDDAD0894DB404482

[84] Record, Jeffrey, *Bounding the Global war on Terror*, Dec 2003, Strategic Studies Institute; *On Point The United States Army in Operation Iraqi Freedom*, Center for Army Lessons Learned, http//onpoint.Leavenworth.army.mil/

[85] *Military Power, Explaining Victory and Defeat in Modern Battle,* Stephen Biddle, Princeton University Press, 2004 pp 1–5

[86] *On Point The United States Army in Operation Iraqi Freedom*, Center for Army Lessons Learned, http//onpoint.Leavenworth.army.mil/; Burgess, Lisa; *Army Program offers Jobs in Three New Unit's of Action,* Stars and Stripes European Edition, http//www.military.com/newconent/0%2C13190%2CSS_050504_Army%2C00.html; **Go**le, Henry G**.** *The Road to Rainbow***:** *Army Planning for Global War, 1934–1940, Naval Institute Press, 2002*

[87] See Colin S Gray, *How has War Changed since the End of the Cold War?,* Parameters, Spring 2005, pp 14–26 for an excellent paper on that comparison and please refer to McNair Paper 62 for a generally accepted definition of Asymmetric warfare

[88] On Strategy, Col Harry Summers; See Colin S Gray, *How has War Changed since the End of the Cold War?,* Parameters, Spring 2005, p5

[89] Antulio Echevarria II, *Toward an AmericanWay ofWar* J. (Carlisle, Pa.: US ArmyWar College, Strategic

Studies Institute, March 2004), p. 1.

[90] Colin S Gray, *How has War Changed since the End of the Cold War?,* Parameters, Spring 2005, p16

[91] General Wesley K Clark, *Waging Modern War*, BBS, 2001, pp 419–429

[92] BG T.R.Phillips, *Roots of Strategy, The Five Greatest Military Classics of All Time*, Sun Tzu, p24

[93] Mao Tse Tung, *Problems of War and Strategy, Peking,* Foreign Language Press, 1960; Mao Tse Tung . *On The Protracted War*

[94] Web article, Defense Link, *Officials discuss Global Posture Process,* http://www.defenselink.mil/news/Jun2004/n06092004_200406097.html See also BASIC NOTES, The *US Global Posture Review: Reshaping America's Global Military Footprint*, http://www.basicint.org/pubs/Notes/BN041119.htm

[95] 1–4-2–1 defense policy: Defend the US; Be capable of Deterring aggression in 4 critical regions; Defeat this aggression in 2 regions simultaneously; Win decisively to include Regime Change in one of those conflicts at a "time and place of our choosing".

Current Pentagon and Administration arrogance is appalling and glaringly obvious in the fact that "under the Rumsfeld strategy the new plans will be focused on "theaterwide" rather than "country-specific" scenarios.[11] It was felt that it would be possible to reduce the number of plans since the view within the Pentagon is that U.S. military forces are so good that they can beat virtually any nation with the same plan and thus do not have to focus on country specific details"

[96] British American Security Information Council, BASIC NOTES, The *US Global Posture Review: Reshaping America's Global Military Footprint*, http://www.basicint.org/pubs/Notes/BN041119.htm

See also footnote number 5

[97] The term Emerging Doctrine as used by the US Army is a synthesis of the doctrinal concepts outlined in FM1, The Army, and FM 3–0, Operations. Emerging doctrine is both the new concept of Asymmetrical warfare and Full Spectrum operations with an added emphasis on operations other than war, asymmetry and Coalition/Multi- National operations

[98] The baseline reference is FM 3–0, Operations. The definition infers and implies that Technology is the decisive factor in asymmetry

[99] The definition in Mc Nair Paper 62 appears to be a better definition as compared to current US Army Doctrine.

[100] Napoleons operation in Egypt suffered from nuisance skirmishes and at Acre, the Plague. Although successful at defeating the Turks at Aboukir Bay on July 25, the entire operation was a strategic debacle

[101] Napoleon had secretly readied Frigates in July for his return to Paris. He had received his 2nd recall notice while at Acre.

[102] These three operations had two essential effects. The first was to put French troops in someone else's territory because the Directory could not sustain them, and they were attempting to prevent the inevitable invasion of France that appeared to be imminent.

[103] Bonaparte is still consolidating his power. The Vendee was in revolt. Additionally, this is considered some of Napoleons most productive and creative time. He re wrote the constitution, the banking system and began overhauling the Army.

[104] A Military History and Atlas of the Napoleonic Wars, Revised Edition, BG Vincent J. Esposito and Col John R. Elting, Greenhill Books, 1999. The French Army Corp consists of 2 to 4 Divisions depending on Task Organization, it consists of a Brigade or division of Light Cavalry, a company or two of Artillery and Engineers and service and support troops, Baggage trains were driven by paid teamsters

[105] Clausewitz was about 15 when the Peace of Basel was signed. His comments about the French Revolution are found in Book 8, On War

[106] The troops were billeted in a wide area by default rather than by design. There was lack of logistics in everything.

[107] Moreau was disobedient, will full and stubborn; by all rights he should have been sacked.

[108] There was still some debate in the planning phase in which of the passes would be the primary pass.

[109] A Military History and Atlas of the Napoleonic Wars, Revised Edition, BG Vincent Esposito and Col John Elting, Stackpole Books, 1999, page/Map 35

[110] Bonaparte knew that part of his success was going to be determined by Mas-

sena and Suchet and how long they could keep the Austrians diverted away from him or the Rhine

[111] By this time, Napoleon had no intention of relieving Massena, but in cutting Melas' lines of communication and getting into his rear area.

[112] French troops are returned to France and Massena promptly boards a frigate and returns to the fray.

[113] Melas had never intended to pull back across the creek into Alessandria. He had intended to use the space to deploy from the line. Inept and scared commanders prevented Melas from gaining the maneuver space and freedom of action he desperately needed.

[114] Napoleon was ruthless in the manner in which he treated ineptitude on this occasion. Monnier and Chambarlhac were removed from the active rolls and Duvignau was made a scapegoat and forced to retire

[115] After Kaim was placed in charge, it appears as though he passed the task of follow up and pursuit to Zach, the Chief of Staff. But the whole affair was not coordinated. To make matters worse, most of the commanders and units observed Melas leaving the battlefield. This added to the confusion.

[116] Desaix had literally marched cross-country to the sound of the Guns. He had deliberately slowed his pace when given the task of screening south. Lucky for Bonaparte that Desaix was somewhat insubordinate in the execution of his duties.

[117] The French were still outnumbered. The Austrians had been defeated psychologically. French Casualties were reported as about 5,400 vs the Austrian 9400. The casualty reports however are suspect.

[118] The comparison is used to show time distance and space with in context of the operation. In another bit of trivia, when Bonaparte left Paris for Geneva on 5 May, he arrived on the 8 or 9th. That was over 240 miles by Horseback with an average daily rate on horse at 60 miles a day. That is *incredible* stamina.

[119] This is from Baron Von der Goltz , the Conduct of War and portrays 4 vari-

ables in tabular form in which to wage war. from, On Strategy, a Critical Analysis of the Vietnam War by Col Summers

[120] I would imagine that Desaix had taught Bonaparte this lesson

[121] . Lapoype was engaged in re crossing the swollen and engorged Po when he received his recall instructions. It had been raining heavily and the terrain was a mess. Lapoype arrived to late to be a part of the battle. I wonder what he *"Really"* said when he received Napoleons recall orders????

[122] Modern US Army concepts of Emerging Doctrine include Asymmetrical warfare and Full Spectrum Operations. In many Way's Napoleons constant task reorganization, logistics of living off the land, maneuver and tactics of penetrating rear area's to sever lines of Communication was the emerging doctrine for the 1800's.

[123] Stalingrad to Berlin: The German Defeat in the East, page 17

[124] This has been expressed in several publications that are freely and widely available on line such as: Sharp Corners: Urban Operations at the Century's End, CSI; The New Urban Battlefield, Newsweek; By The Edge of The Sword: A Consideration of the Challenges Inherent in Modern Urban Military Operations and several others listed in the Bibliography

[125] recently released draft that is intended to replace FM 90–10 and FM 90–10–1 a pdf copy of this document is available from the TRADOC, General Dennis Reimer on line library or from the Collaborate section of AKO under the "CADD" community

[126] These manuals and documents provide the strategic, operational and tactical basis for training and preparing the Military for its Missions and ability to achieve Full Spectrum Dominance. Asymmetry is but one element.

[127] Page 81 of Stalingrad by Antony Beevor, Field Marshall Gerd Von Rundstedt may have been correct in his assessment of the "Coalition" Yet, I would surmise that his public expressions also had a lot to do with how he also helped "Create" that climate within his Officer Corp.

[128] Ibid, pg86,180,182,279,286,307–308,319,322,350,378

[129] Ibid, the author describes numerous atrocities, of particular mention is page 14 and 15 and Chapter 11

[130] Ibid, Chapter 11

[131] Article in the New York Times, July 13, 1986, The Split Military Psyche, talks about the bitter inter-service rivalry and power politics which center on budgets, resources, and manpower. The implication was the same in 42 between all the elements of the Nazi Military, especially between the Army and the Luftwaffe

[132] Ibid 5, pages 148–150

[133] This is in reference to Von Clausewitz, On War, Book 1, Chapter 7, ed Michael Howard and peter Paret

[134] Ibid 5, Chapter 25 page 422–431

[135] McNair Paper 62 offers a better definition of Asymmetry as opposed to the Doctrinal Definition in FM 3–0, Operations, Chapter 4

[136] NIV, Zondervan, Judges 14:4

[137] Concordia Self Study Bible with Bible Atlas, NIV, St Louis, ed, Robert G Hoerber, 1984

[138] Business week , dated Jan 20 2003 Please See "Foreign Policy: Bush's new Pragmatism" a Commentary by Bruce Nussbaum on page 38 of the section "News: Analysis and Commentary"

[139] Current doctrine on training and evaluation, FM 7–0, Train the Force, implies as much by directing that evaluations are conducted and coached two levels down; for instance a Battalion Commander trains and evaluates Company Commanders, but also mentors Platoon Leaders.

[140] Page 673 from American Military History, US Army Center for Military History, 1989

[141] Pages 43–45; The Art and Practice of Military Strategy, National Defense University, 1984

[142] page 110, table I, On Strategy, Col Harry G. Summers.

[143] Ibid (1) Page 672

[144] Ibid page 10; 3 battalions of the 4th NVA Regiment, 3 Battalions of the 6th NVA Regiment, an Independent Battalion and 2 sapper battalions

[145] Ibid page 68–69

[146] Ibid page 68–69, 78, This refers to both the Call for Fire support Debacle; when it took the chain of command 2 hours to deny a fire mission. And the MACV Col that demanded that the American Flag be taken Down from the Capitol Building

[147] Page 1 and 2 of Harry Summers Book "On Strategy

[148] From the American Way of War by Weigley, page 457

[149] from the CSI Essay, Compound Warfare, by Randall Briggs

[150] Reference to Madeline Albright and UNOSOM II

[151] UNOSOM II Mission statement and goals, The following website has a direct link to that document. http://www.geocities.com/CapitolHill/8514/somalia.htm

See all the following website from PBS, Frontline. They have an excellent chronology of events. http://www.pbs.org/wgbh/pages/frontline/shows/ambush/

[152] Refer to this website for the Political-Strategic analysis that refers to this. See also the book Black Hawk Down, page 71–72 http://www.geocities.com/CapitolHill/8514/somalia.htm

[153] refer to the same website above for the rest and bounty and also the book Black Hawk Down

[154] Ibid

[155] Tier One personalities are found on page 8, Black Hawk Down

[156] This is a reference to Clausewitz, On War

[157] Asymmetrical definitions in FM 3–0 emphasize the technological disparities. McNair paper 62 offers a better definition.

[158] This is a reference to the National Security Strategy and Bibliography reference number 5 and 6

[159] *Sharp Corners: Urban Operations at Century's End*, Roger J. Spiller, US Army Command and General Staff College Press; *Block by Block: The Challenge of Urban Operations*, William G Robertson, Lawrence A. Yates, US Army Command and General Staff College Press.

[160] Population Resource Center; prc@prcnj.org.

161 Forgotten Armies, The Fall of British Asia, 1941–1945, Christopher Bayly and Tim Harper, Harvard University Press, 2006

BIBLIOGRAPHY FOR PROLOGUE TO CURRENT DOCTRINE AND ASYMMETRICAL LESSONS

On War, Clausewitz, ed by M. Howard and P Paret, Princeton, 1976

The Art of War, Sun Tzu, ed and tr, Samuel Griffith, Oxford, 1971

the Book of Five Rings, Miyamato Musashi

On War, Frederick the Great, ed by Jay Luvaas, New York, 1966

Roots of Strategy, Book 4,ed David Jablonsky, Stackpole 1999

Historical and Political Writings, Carl Von Clausewitz, ed. and tr. By Peter Paret and Daniel Moran, Princeton, 1992

Clausewitz and the State, Paret , Oxford, 1976

Makers of Modern Strategy from Machiavelli to the Nuclear Age, Petet Paret, Princeton

Makers of Modern Strategy Military thought from Machiavelli to Hitler, Edward Mead Earle, Princeton;

The Prince, Nicolo Machiavelli, translated by W.K.Marriot,Macmillian,1915

McNair Paper 62, *The Revenge of the Melians: Asymmetric Threats and the next QDR*, November 2000

A National Security Strategy for the New World Coming, The White House, December 1999

New World Coming: American Security in the 21st Century, Phase I report of the US Commission on National Security /21st Century, September 1999

Roadmap for National Security: Imperatives for Change, Phase III Report of the US Commission on National Security / 21st Century, February 2001

Seeking a National Strategy: A Concert for Preserving Security and Promoting Freedom, a Phase II Report of the US Commission on National Security /21st Century, April 2000

New World Coming: American Security in the 21st Century: Supporting Research and Analysis, September 1999

Quadrennial Defense Review Report, US Dept of Defense, Sept 2001

Report of the National Defense University Quadrennial Defense Review 2001 Working Group

Asymmetry and US Military Strategy: Definitions, Background and Strategic Concepts, Steven Metz and Douglas V Johnson II, Jan 2001

Field Manual 3–0 Operations, US Army, 2001

A concise Dictionary of Military Biography the careers and campaigns of 200 of the most Important Military Leaders, Martin Windrow and Francis K mason, John Wiley and Sons

Brassey's Encyclopedia of Military History and Biography, Brassey's , Franklin D Margiotta, 2000

Brassey's Encyclopedia of Land Warfare, Brassey's, Franklin D Margiotta, 2000

The Art and Practice of Military Strategy , National Defense University, George E. Thibault, 1984

Webster's Collegiate Dictionary, 10th Edition

Small Unit Actions During the German Campaign in Russia, CMH Pub 104–22, US Army Center for Military History, 1988

The "Love Bug" and Computer Attacks; Future National Security Implications... by C. L. Staten, CEO and Sr. Analyst, Emergency Response & Research Institute (ERRI)

The Army in the Strategic Planning Process: Who Shall Bell the Cat, Bethesda; concepts analysis report, Carl H. Builder, 1989

On Strategy, a critical analysis of the Vietnam War, Harry G. Summers Jr. Presidio, 1982 The Landmark Thucydides: a comprehensive guide to the Peloponnesian war; ed, Robert B. Strassler; Simon & S huster; 1996

Thucydides on the Nature of Power, A Geoffrey Woodhead, Harvard, 1970 (University of New England, Armidale, New South Wales, Australia, Call Number 888.2/zw888)

Thuycidides Narrative and Explanation, Tim Rood, Oxford, 1998(University of New England, Armidale, New South Wales, Australia, Call Number 888.2/R776)

Thucydides and Athenian Imperialism, Jacqueline De Romilly, Professor of Greek at the Sorbonne, Oxford, 1963, (University of New England, Armidale, New South Wales, Australia, Call Number, 888.2/ZR765)

The Art of War in World History: From Antiquity to the Nuclear Age; Gerard Chaliand; Univ of Calif Press; 1994

The Art of War in the Western World; Archer Jones; Oxford; 1987

The History of the Art of War Within the Framework of Political History; Hans Delbruck; Vol 1, Antiquity; Tr, Walter J. Renfroe, Jr. Greenwood Press, 1975

The Greek State at War, Volumes I to V, W.Kendrick Pritchett,FBA; University of California Press, 1971

The Making of Strategy: Rulers, states and war; Murray, Knox and Bernstein; Cambridge; 1994

The Encyclopedia of Military History from 3500 b.c. to present; Dupuy and Dupuy; Harper and Row; 1986

McNair Paper 62, The Revenge of the Melians: Asymmetric Threats and the next QDR, November 2000

Contemporary Operational Environment, The Center for Army Lessons Learned, Col Michael Heimstra, TRADOC, 2002

Defining and Achieving Decisive Victory; Colin S. Gray; Strategic Studies Institute; 2002

Roadmap for National Security: Imperatives for Change, Phase III Report of the US Commission on National Security / 21st Century, February 2001

Asymmetry and US Military Strategy: Definitions, Background and Strategic Concepts, Steven Metz and Douglas V Johnson II, Jan 2001

Brassey's Encyclopedia of Military History and Biography, Brassey's , Franklin D Margiotta, 2000

Brassey's Encyclopedia of Land Warfare, Brassey's, Franklin D Margiotta, 2000

The Art and Practice of Military Strategy , National Defense University, George E. Thibault, 1984

Field Manual 3–0 Operations, US Army, 2001

Field Manual 1, The Army, 2001

On War, Clausewitz, ed by M. Howard and P Paret, Princeton, 1976

BIBLIOGRAPHY FOR MARENGO CAMPAIGN: ASYMMETRICAL ANALYSIS

1. The Encyclopedia of Military History from 3500bc to the present, 2nd Revised Edition, R. Ernest Dupuy and Trevor R. Dupuy, Harper and Row, 1986

2. Brassey's Encyclopedia of Military History and Biography, ed, Franklin D Margiotta, Brassey's, 2000

3. A Military History and Atlas of the Napoleonic Wars, Revised Edition, BG Vincent J. Esposito and Col John R. Elting, Greenhill Books, 1999

4. A Military History and Atlas of the Napoleonic Wars, Revised Edition, BG Vincent J. Esposito and Col John R. Elting, Preager, 1964

5. On Strategy, a critical analysis of the Vietnam War, Harry G. Summers Jr. Presidio, 1982

6. Field Manual 3–0 Operations, US Army, 2001

7. Asymmetry and US Military Strategy: Definitions, Background and Strategic Concepts, Steven Metz and Douglas V Johnson II, Jan 2001 McNair Paper 62, *The Revenge of the Melians: Asymmetric Threats and the next QDR*, November 2000

8. Clausewitz and the State, Paret , Oxford, 1976

   On War, Clausewitz, ed by M. Howard and P Paret, Princeton, 1976

9. The Campaigns of Napoleon, David G. Chandler, Scribner, 1966

10. Soldiers at War Firsthand accounts of warfare from the age of Napoleon; With Napoleon in Italy, Capt Jean-Roch Coignet, 96th Demi Brigade, ed, Jon E. Lewis, Carroll and Graf, 2001

11. Napoleon's Great Adversaries: The Archduke Charles and the Austrian Army 1792–1814, Indiana University Press, 1982

12. Warfare in the Western World, Vol I, Military operations from 1600–1871, Doughty, Gruber, D.C. Heath, 1996

13. The Art of War in World History, Gerard Chaliand, University of California, 1994

14. The Makers of Modern Strategy from Machiavelli to the Nuclear age, ed, Peter Paret, Princeton, 1986

15. The Napoleonic Wars, Gunther Rothenberg, Cassell, 1999

16. Dictionary of Military and Naval Quotations, Robert Debs Heinl, US Naval Institute, 1966

17. Source Book of the Marengo Campaign in 1800, The General Service Schools, The General Staff School, The General Service Schools Press, ed Conrad Lanza, 1922

18. The Campaign of Marengo, Herbert H Sergeant, Chicago, 1897

19. Patterns of War since the Eighteenth Century, 2nd Edition, Indiana University Press, Larry H Addington, 1994

20. The Art of Warfare in the Age of Napoleon, Indiana University Press, Gunther E. Rothenberg, 1980

BIBLIOGRAPHY FOR THREE CITIES ASYMMETRICAL ANALYSIS: STALINGRAD, MOGADISHU, HUE

Stalingrad

Stalingrad The Fateful Siege: 1942–1943, Penguin, 1998, Antony Beevor

Stalingrad to Berlin: The German Defeat in the East, Army Historical Series, US Government Printing Office ,1971, Earl F. Ziemke,

The Encyclopedia Of Military History from 3500 b.c. to the present, Harper and Row, 1986, R.Ernest Dupuy and Trevor N. Dupuy.

New World Coming American Security in the 21ST Century: Major Themes and Implications, Phase I report on the emerging Global Security Environment for the first Quarter of the 21st Century, US Commission on National Security, September 15, 1999.

Quadrennial Defense Review Report, September 30th 2001

FM-3–06.11 (Draft) Combined Arms Operations in Urban Terrain, US Army, Undated

FM 3–0, Operations, US Army, June 2001

Sharp Corners Urban Operations at Century's End, CSI, Rodger J Spiller

The Battle of Stalingrad, CSI, S.J.Lewis

The New Urban Battlefield U.S troops don't do cities. But someday they'll have to, Newsweek, February 21 2000

By The Edge of The Sword: A consideration of the Challenges Inherent in Modern Urban Operations, YALE, Peter J. Gulliver

Attacking the Heart and Guts: Urban Operations through the Ages, CSI, Lou DiMarco

Conference Summary Urban Warfare" Options, Problems and the Future, 1999, Daryl G. Press

The Human Terrain of Urban Operations, Ralph Peters, Parameters, Spring 2000

Urban Combat: Confronting the Specter, CALL, Lester W. Grau and Dr. Jacob W. Kipp, Foreign Military Studies Office, Ft Leavenworth, KS, Military Review, July-Aug 1999

The Split Military Psyche, Arthur T Hadley, July 13 1986, the New York Times.

ON War, Carl Von Clausewitz, ed, Michael Howard and Peter Paret, Princeton, 1976

McNair Paper 62; The Revenge of the Melians: Asymmetric Threats & the Next QDR, Nov 2000

Business Week Magazine, dated Jan 20 2003, page 38

Concordia Self Study Bible with Bible Atlas, NIV, St Louis, ed, Robert G Hoerber, 1984

Soldiers in Cities: Military Operations on Urban Terrain, ed, Michael C. Desch, Oct 2001 (A Compendium of Historical and Contemporary Essays)

Battle for Hue Tet 1986, Presidio, Keith William Nolan, 1996

The Encyclopedia of Military History from 3500 b.c. to the present, Harper and Row, 1986, R.Ernest Dupuy and Trevor N. Dupuy.

New World Coming American Security in the 21ST Century: Major Themes and Implications, Phase I report on the emerging Global Security Environment for the first Quarter of the 21st Century, US Commission on National Security, September 15, 1999.

Quadrennial Defense Review Report, September 30th 2001

FM-3–06.11,Combined Arms Operations in Urban Terrain, US Army, February 28, 2002

FM 3–0, Operations, US Army, June 2001

Sharp Corners Urban Operations at Century's End, CSI, Rodger J Spiller

The Battle for Hue, CSI, James H. Wilbanks

By The Edge of The Sword: A consideration of the Challenges Inherent in Modern Urban Operations, YALE, Peter J. Gulliver

Attacking the Heart and Guts: Urban Operations through the Ages, CSI, Lou DiMarco

Conference Summary Urban Warfare" Options, Problems and the Future, 1999, Daryl G. Press

The Human Terrain of Urban Operations, Ralph Peters, Parameters, Spring 2000

Urban Combat: Confronting the Specter, CALL, Lester W. Grau and Dr. Jacob W. Kipp, Foreign Military Studies Office, Ft Leavenworth, KS, Military Review, July-Aug 1999

ON War, Carl Von Clausewitz, ed, Michael Howard and Peter Paret, Princeton, 1976

McNair Paper 62; The Revenge of the Melians: Asymmetric Threats & the Next QDR, Nov 2000

Soldiers in Cities: Military Operations on Urban Terrain, ed, Michael C. Desch, Oct 2001 (A Compendium of Historical and Contemporary Essays) pages 75–87 and pages 149–166

On Strategy a critical analysis of the Vietnam War, Harry G. Summers Jr.; Presidio, 1995

American Military History, Center of Military History, 1989

The Art and Practice of Military Strategy, ed George Edward Thibault, National Defense University, 1984

The Army and Vietnam, Andrew F. Krepinevich,Jr.; John Hopkins University Press, 1986

The American Way of War a history of United States Military Strategy and Policy, Russell F. Weigley, Indiana University Press, 1973

Compound Warfare an Anthology of Combat Studies, CSI, Randall Briggs

Black Hawk Down a story of modern war, Penguin,,1999, Mark Bowden

Somalia Operations, Kenneth Allard, National Defense University Press, 1995

Somalia, UNOSOM II, Department of Public Information, United Nations, 1997

The Encyclopedia of Military History from 3500 b.c. to the present, Harper and Row, 1986, R.Ernest Dupuy and Trevor N. Dupuy.

New World Coming American Security in the 21ST Century: Major Themes and Implications, Phase I report on the emerging Global Security Environment for the first Quarter of the 21st Century, US Commission on National Security, September 15, 1999.

Quadrennial Defense Review Report, September 30th 2001

FM-3–06.11 Combined Arms Operations in Urban Terrain, US Army, February 28, 2002

FM 3–0, Operations, US Army, June 2001

Sharp Corners Urban Operations at Century's End, CSI, Rodger J Spiller

By The Edge of The Sword: A consideration of the Challenges Inherent in Modern Urban Operations, YALE, Peter J. Gulliver

Attacking the Heart and Guts: Urban Operations through the Ages, CSI, Lou DiMarco

Conference Summary Urban Warfare" Options, Problems and the Future, 1999, Daryl G. Press

The Human Terrain of Urban Operations, Ralph Peters, Parameters, Spring 2000

Urban Combat: Confronting the Specter, CALL, Lester W. Grau and Dr. Jacob W. Kipp, Foreign Military Studies Office, Ft Leavenworth, KS, Military Review, July-Aug 1999

ON War, Carl Von Clausewitz, ed, Michael Howard and Peter Paret, Princeton, 1976

McNair Paper 62; The Revenge of the Melians: Asymmetric Threats & the Next QDR, Nov 2000

Soldiers in Cities: Military Operations on Urban Terrain, ed, Michael C. Desch, Oct 2001 (A Compendium of Historical and Contemporary Essays)

On Strategy a critical analysis of the Vietnam War, Harry G. Summers Jr.; Presidio, 1995

American Military History, Center of Military History, 1989

The Art and Practice of Military Strategy, ed George Edward Thibault, National Defense University, 1984

The American Way of War a history of United States Military Strategy and Policy, Russell F. Weigley, Indiana University Press, 1973

Compound Warfare an Anthology of Combat Studies, CSI, Randall Briggs

Block by Block: The Challenge of Urban Operations, William G Robertson, Lawrence A. Yates, US Army Command and General Staff College Press.

Forgotten Armies, The Fall of British Asia, 1941–1945, Christopher Bayly and Tim Harper, Harvard University Press, 2006

GLOSSARY

ANA	Afghan National Army
ARVN	Army Vietnam
BCT	Brigade Combat Team
C3	Command, Control, Communications
CALL	Center For Army Lessons learned
CENTCOM	Central Command

COIN	Counterinsurgency
CFC-A	Combined Forces Command-Afghanistan
CSS	Combat Service Support
CSTC-A	Combined Security Transition Command-Afghanistan
ETT	Embedded Training Teams
FM	Field Manual
GWOT	Global War on Terror
HVT	High Value Target
IPB	Intelligence Preparation of the Battlefield
JTOC	Joint Tactical Operations Center
JRTC	Joint Readiness Training Center

JV	Joint Vision
LOG	Logistics
LZ	Landing Zone
MACV	Military Assistance Command Vietnam
MDMP	Military Decision Making Process
MOUT	Military Operations Urban Terrain
MPRI	Military Professional Resources Inc.
MVT	Medium Value Target
NATO	North Atlantic Treaty Organization
NOD	Night Observation Device
NCO	Non-Commissioned Officer

NDU	National Defense University
NSS	National Security Strategy
NTC	National Training Center
NVA	North Vietnamese Army
OOTW	Operations Other Than War
RMA	Revolution in Military Affairs
ROE/ROI	Rules of Engagement/Interaction
QDR	Quadrennial Defense Review
SASO	Security and Stability Operations
SOCOM	Special Operations Command
SOF	Special Operations Forces

SOP	Standard Operating Procedures
SWAT	Special Weapons and Tactics
TAG	Training and Assistance Group
TO&E	Table of Organization and Equipment
TRADOC	Training And Doctrine Command
TTP	Tactics, Techniques and Procedures
UA	Unit of Action
UE	Unit of Employment
UN	United Nations
USJFCOM	United States Joint Forces Command